T0350895

Fatigue in Aviation

This updated edition includes fatigue and sleep definitions as well as strategies for the measurement and assessment of fatigue. The aviation performance, mood, and safety problems associated with sleep restriction and circadian disruptions in operational settings are highlighted. The biological bases of fatigue are discussed so that the reader can understand that it is a real physiological phenomenon and not 'just a state of mind'. Both traditional and newly-developed scientifically-valid countermeasures are presented, and a variety of data from diverse sources are included to provide readers with a 'toolbox' from which they can choose the best solutions for the fatigue-related problems that exist in their unique operational context. In addition, an essential overview of Fatigue Risk Management Systems is included to provide the basic structure necessary to build and validate a modern, integrated approach to successful fatigue management.

The book is of interest to aviation crews in both civilian and military sectors – managers as well as pilots, flight crews, and maintainers. It aims to be user-friendly, although scientific information is included to help the reader fully understand the 'fatigue phenomenon' from an evidence-based perspective as well as to enhance the reader's appreciation for the manner in which various counter-fatigue interventions are helpful.

Dr. John A. Caldwell has a Ph.D. in Experimental Psychology with over 25 years of experience conducting multidisciplinary research in settings including the U.S. Army's Medical Research and Materiel Command, NASA, and the U.S. Air Force Research Laboratory. He is an internationally recognized expert scientist in the areas of sleep deprivation, fatigue countermeasures, and research on the real-world performance effects of select pharmacological compounds. He is a Fellow in the Aerospace Medical Association and the Aerospace Human Factors Association. In addition to developing and conducting a variety of workshops, Dr. Caldwell has made over 100 scientific and programmatic presentations, published numerous book chapters, peer-reviewed scientific papers, articles and government reports and provided over 60 national television/radio/print interviews. Among his awards, Dr. Caldwell has received the Air Force's highest civilian award for research and development.

Dr. J. Lynn Caldwell has a Ph.D. in Experimental Psychology with over 20 years of experience investigating sleep-optimization and fatigue-mitigation strategies in military aviation. She is certified by the American Board of Sleep Medicine as a Sleep Specialist. In her work with the U.S. Army, U.S. Air Force, and U.S. Navy she has conducted numerous simulator and in-flight investigations on fatigue countermeasures, sleep, circadian rhythms, and pharmacological interventions in rated military pilots. She has served as a Distinguished Visiting Scholar at the U.S. Air Force Academy, and she is a Fellow of the American Academy of Sleep Medicine, the Aerospace Medical Association, and the Aerospace Human Factors Association. She serves as an internationally-recognized fatigue-management consultant for a variety of military and civilian groups. She has published extensively in the peer-reviewed literature, and she frequently conducts workshops, safety briefings, and training courses for aviation personnel, flight surgeons, commanders, and safety officers.

Fatigue in Aviation

A Guide to Staying Awake at the Stick

2nd Edition

John A. Caldwell
and
J. Lynn Caldwell

LONDON AND NEW YORK

First published 2016
by Routledge
2 Park Square, Milton Park, Abingdon, Oxon OX14 4RN

and by Routledge
605 Third Avenue, New York, NY 10017

Routledge is an imprint of the Taylor & Francis Group, an informa business

British Library Cataloguing in Publication Data
A catalogue record for this book is available from the British Library

Library of Congress Cataloguing in Publication Data
Names: Caldwell, John A., author. | Caldwell, J. Lynn, 1958– author.
Title: Fatigue in aviation : a guide to staying awake at the stick /
by John A. Caldwell, Jr. and J. Lynn Caldwell.
Description: Burlington, VT : Ashgate, [2016] | "This updated edition includes fatigue and sleep definitions as well as strategies for the measurement and assessment of fatigue. The aviation performance, mood, and safety problems associated with sleep restriction and circadian disruptions in operational settings are highlighted. The biological bases of fatigue are discussed so that the reader can understand that it is a real physiological phenomenon and not 'just a state of mind'. Both traditional and newly-developed scientifically-valid countermeasures are presented, and a variety of data from diverse sources are included to provide readers with a 'toolbox' from which they can choose the best solutions for the fatigue-related problems that exist in their unique operational context. In addition, an essential overview of Fatigue Risk Management Systems is included to provide the basic structure necessary to build and validate a modern, integrated approach to successful fatigue management" – Provided by publisher. |
Includes bibliographical references and index.
Identifiers: LCCN 2015036446 | ISBN 9781472464590 (hardback : alk. paper) | ISBN 9781315582030 (ebook) | ISBN 9781317136231 (epub)
Subjects: LCSH: Air pilots–Health and hygiene. | Fatigue. |
Sleep disorders. | Aviation psychology.
Classification: LCC RC1085.C355 2016 | DDC 616.9/80213–dc23
LC record available at http://lccn.loc.gov/2015036446

ISBN 13: 978-1-4724-6459-0 (hbk)
ISBN 13: 978-1-315-58203-0 (ebk)
ISBN 13: 978-1-317-13624-8 (web PDF)
ISBN 13: 978-1-317-13623-1 (ePub)
ISBN 13: 978-1-317-13622-4 (mobi/kindle)

Typeset in Sabon
by Out of House Publishing

Contents

List of Figures

List of Tables

Part I

The Problem of Fatigue

1 Introduction

Fatigue has long been a concern in modern aviation. However, in 2009 the crash of Colgan Air Flight 3407 brought it to the forefront after this mishap triggered a wave of questions about the commuting practices and duty schedules associated with commercial flight operations in the US. Suddenly it seemed that "doing something about overly tired pilots" was a major topic of conversation even though for many years prior to this incident it looked like revisions to crew scheduling guidelines were unlikely. This was despite the fact that fatigue had long been on the National Transportation Safety Board's (NTSB) Top 10 Most Wanted List of needed safety improvements.

Is aviator fatigue a genuine concern, or is it just a great topic for media hype? Well, according to pilots like Captain Stephen Dodge, fatigue is now and always has been a big problem for both military and civilian aircrews. After completing two Tonkin Gulf cruises flying anti-submarine warfare missions off the USS Bennington, Steve thought he had pretty much become accustomed to the "24/7," rotating schedules common in military aviation. However, after joining Pan American World Airways in 1968, he was surprised to find out just how tiring multi-leg, multi-time zone, long-range commercial operations could be. His ten-day trips outbound from San Francisco with layovers in Honolulu, Tokyo, and Hong Kong, were routinely followed by a four-leg day in and out of Viet Nam (twice), a layover in Taipei, and then another four-leg day in and out of Viet Nam. Afterwards, there was the additional Hong Kong layover en route to Tokyo, where (after everyone was adjusted to Western Pacific Time) they made an afternoon departure for the non-stop night flight back to San Francisco. According to Steve, nothing before or since has topped the fatigue he felt while his head bobbed and his burning eyes squinted into the rising sun on that final ten-hour, eastbound leg across the Pacific. And keep in mind that since then he has dealt with plenty of six-day legs dodging thunderstorms on the Boston–La Guardia-Washington shuttle, numerous transatlantic flights with "tag" flights after all-nighters across the ocean, and an abundance of six-leg days to and from Berlin with multiple Category II approaches to minimums! In his opinion, tired pilots deserve every bit of the attention they receive.

Steve's concerns are echoed by military pilots like Colonel Gary Woltering. In 1996, Gary led eight US Air Force F-117 stealth fighters on a record non-stop mission from Holloman Air Force Base, New Mexico, to an initially undisclosed destination in the Persian Gulf region. Poor weather made the late-afternoon takeoff, the in-flight refuels from the KC-10s, and the outbound trip over the Atlantic more difficult than usual, but uncertainty about the final destination added stresses of its own. It wasn't until the group crossed over the Rock of Gibraltar that Gary finally got the word on where they were going. Then, after digging through 3 feet of pubs to find the Al Jaber approach plates, he flew across Egypt through intermediate ceilings before starting the approach into Kuwait. By now, everyone was low on fuel and the effect of being strapped into an ejection seat for 18 straight hours was taking its toll. However, the situation was soon compounded by a 35-knot crosswind, swirling desert sand, and below-minimum conditions for the instrument approach. Realizing he had to get his planes on the ground before the aircraft (and the pilots) all ran out of gas, Gary fought through the haze of fatigue and devised a plan to split for the Instrument Landing System (ILS) and use the IRADS (Infra Red Acquisition and Detection System – a bombing sensor, not an instrument approach aid) to find the runway, make the landing, deploy the drag chutes, and hit the brakes. When he announced this novel strategy to the rest of the guys, he sounded a lot more confident than he actually felt, but despite his anxiety, there was no other realistic choice. As soon as he landed, he immediately positioned his jet so he could see the final approach course, and he watched breathlessly as each of his 117s appeared out of the dust and made it safely to the ground. To this day, Gary will never forget his struggles to think clearly enough to get the job done despite the severe fatigue that this grueling, but strategically necessary, mission produced. It was all he could do to drag himself out of the cockpit after he was sure everyone was okay. In Gary's opinion, the issue of pilot fatigue is far more than just an interesting topic for the evening news.

How many times have you struggled through those seemingly endless days when the forces of nature, maintenance delays, and miserable schedules left you to wonder how in the world you would ever make it through the flight, much less through the drive home or to the hotel or the base lodging facility? How often have the scratchy eyes, those "out-of-focus" instruments, the head-bobs, and those really annoying heavy eyelids (just checking for pinholes of course) made it clear that the alertness of only a few hours ago was definitely a thing of the past? How many times has that dreaded alarm clock or wakeup call crashed into the peacefulness of your all-too-short layover or pre-mission sleep, forcing you out of bed and into a shower (if you were lucky) that was not nearly as long as the one you needed to have before you rushed to grab those two extra-large cups of coffee on the trip back to the flight line? What's the big deal? This is just the way it is when the job involves the constant challenge of forcing uncontrollable factors into a well-defined, man-made routine while ensuring the company remains profitable, the people get where they are going when they want to go there, the packages arrive "just in time," and the mission gets done when it's supposed to.

Why are we suddenly hearing so much more about pilot fatigue than ever before? Is aviator fatigue really something to worry about? And even if it is, what are the chances that there are actually solutions that can be practically implemented in the hectic worlds of commercial and military aviation? In the upcoming pages, we will attempt to answer each of these questions in a straightforward fashion. The reasons for focusing on weary crews will be presented as well as explanations of the factors that are at the roots of aircrew fatigue. In addition, we will offer practical advice about fatigue-avoidance techniques for operational settings. Finally, we will provide a general overview of the importance of implementing a solid, proactive program of alertness management in present-day aviation operations. But first, the scientific facts that underlie the modern-day concerns with pilot fatigue will be reviewed.

Academic Proof That Fatigue Really Is an Aviation Concern

Some readers may be skeptical that aircrew fatigue deserves the attention it has been receiving over the past few years; however, the facts of the matter show there are valid reasons for concern. Many of these reasons will be presented shortly. In addition, just in case there are some "doubting Thomas's" who might question the relevance of some of this background material, all of the facts presented in this review include the references for each of the statements made. The full text of these sources can be easily obtained and further evaluated with a quick trip to the local university library (and some can be ordered or viewed via the internet).

Fatigue Drivers in Civil Aviation

The global economy, societal factors, and technological advancements have made the issue of aircrew fatigue more important now than ever before in the civilian aviation community. Prior to the September 11 terrorist attacks on US soil, there was an unprecedented amount of international travel supporting multinational business endeavors in the ever-expanding global economy. As a result, statistics suggested air travel would increase by nearly 5 percent per year for the next 20 years (Boeing, 2001a). To increase capacity, major aircraft manufacturers designed new ultra-long-range jets capable of extended long-haul flights that went well beyond the duration of the traditional routes. Aircraft like the Boeing 777–300ER and the Airbus 380–800 are currently flying non-stop routes of over 8,000 miles and almost 17 hours in duration. New aircraft capabilities are here, and despite the fact that terrorist activity, economic factors, and global unrest once dampened passenger growth forecasts, the downturn was only temporary. At the present time, the 5 percent predicted annual growth in air travel appears solid with the global passenger traffic results for 2014 showing a 5.9 percent increase in demand compared to the full year of 2013 (International Air Transport Association, 2015). Thus 2014 performance was above the ten-year average growth rate, and all indications are that this pattern will be sustained in 2015 and beyond. The flying public's demand for a range of convenient departure and arrival schedules with the ability to travel the longest distances in the shortest possible time, day in and day out, is now a routine fact of life, and pilots are required to cope with the scheduling factors that these types of operations impose.

Fatigue Drivers in the Military Sector

Meanwhile, the military aviation community has also extended the timing, range, and duration of flight missions to unprecedented levels. The refinement of night-vision technologies in combination with the fielding of highly reliable aircraft have made around-the-clock missions a critical and ever-present feature of current US combat strategy. In fact, the ability to fly and fight 24 hours a day is a key strategy for winning battles by forcing the enemy to continuously respond to the demands of combat, for days and nights on end, without the benefit of rest and recuperation. In the military rotary-wing community, extended-range fuel systems have stretched helicopter flight durations from the old two-hour standard to as much as eight continuous hours. In fact, helicopter missions are often further extended by "hot refuels" which permit the refilling of fuel tanks with the engines running and the blades turning, provided that the aviators remain seated at the controls. In the military fixed-wing community, long-range intercontinental flights are essential to "Global Strike" and "Global Mobility" strategies that depend upon the capability to rapidly deploy people and weapons to any point on Earth with barely more than a moment's notice. The B-2 stealth bomber can traverse 6,000 nautical miles on a single fuel load and can be refueled in flight. This aircraft now flies non-stop missions of over 40 hours with an unaugmented crew of only two pilots. While certainly stressing human capabilities, such long-duration flights are key to the persistent and sustained operations 24 hours a day, seven days a week that are considered essential to attaining US military victory in today's conflicts. Both US Air Force and Army doctrine emphasize the tactical necessity of sustained and continuous operations in modern warfare (Department of the Army, 1991; Department of the Air Force, 1997).

Work Pressures and Technological Factors

The advent of long-range aircraft in combination with technologies that are effectively "turning the night into day" have significantly augmented the capabilities of the civil and military aviation communities. In the civilian world, travelers arrive at their destinations faster and more conveniently than ever before, and airlines benefit from increased revenues. In the military, tactical objectives are more effectively accomplished, and the lives of "friendly forces" are better preserved. However, there are costs associated with these benefits. Specifically, ultra-long-range, around-the-clock flights are undoubtedly further stressing the limits of human tolerance, which are already being tested in modern aviation. For some time, pilot fatigue has been a significant concern in flight operations. NASA's Aviation Safety Reporting System (ASRS) routinely receives reports from pilots blaming fatigue, sleep loss, and sleepiness in the cockpit for operational errors such as altitude and course deviations, fuel miscalculations, landings without proper clearances, and landings on incorrect runways (Rosekind et al., 1994). Unfortunately, such reports are not surprising given that cockpits are quite conducive to sleep when flying for long durations at high altitudes, particularly at night (Moore-Ede, 1993); and night operations, in general, are known to be associated with higher rates of errors and accidents (Dinges, 1995).

Aviator Fatigue is a Recognized Problem

Pilot fatigue is an insidious threat throughout aviation, but especially in operations involving sleep loss from circadian disruptions, increased sleep pressure from extended duty, and impaired alertness associated with night work (Akerstedt, 1995a). Aviator fatigue is associated with degradations in response accuracy and speed, the unconscious acceptance of lower standards of performance, impairments in the capacity to integrate information, and narrowing of attention that can lead to forgetting or ignoring important aspects of flight tasks (Perry, 1974). Fatigued pilots tend to decrease their physical activity, withdraw from social interactions, and lose the ability to effectively divide mental resources among different tasks. As sleepiness levels increase, performance becomes less consistent and vigilance deteriorates (Dinges, 1990). Even the most basic types of psychomotor performance are degraded by sleepiness/fatigue. Thus, it is clear that fatigue is a threat to flight safety; however, it has been difficult to establish the full and exact cost of fatigue in terms of incidents and accidents.

Fatigue is a Contributor to Air Mishaps

Sixty-five percent of air accidents have been attributed to human error since the start of the jet age, but the percentage attributable to sleep loss/fatigue remains somewhat uncertain (Lauber and Kayten, 1988). An often-cited NTSB study of major accidents in domestic air carriers from 1978 through 1990 in part concluded that, "… Crews comprising captains and first officers whose time since awakening was above the median for their crew position made more errors overall, and significantly more procedural and tactical decision errors" (NTSB, 1994: p. 75). Kirsch (1996) estimated that fatigue may be involved in 4–7 percent of civil aviation mishaps, and Marcus and Rosekind (2015) noted that among the major NTSB investigations of aviation mishaps between 2001 and 2012, fatigue was noted as a factor in 23 percent. Data from the US Army have suggested fatigue is involved in 4 percent of Army accidents (Caldwell, Gilreath, and Erickson, 2002), and Air Force data from 2006 and 2007, indicates that fatigue was cited in over 20 percent of the Class A aviation mishaps (Musselman, 2008). The US Air Force Safety Center estimates that aircrew fatigue contributes to annual losses of approximately $54 million (personal communication, T. Luna, 2002). Furthermore 25 percent of the Air Force's night tactical fighter Class A accidents were attributed to fatigue between 1974 and 1992, and 12.2 percent of the Navy's total Class A mishaps were thought to be the result of aircrew fatigue from 1977 to 1990 (Ramsey and McGlohn, 1997). At first glance, some of these percentages may seem rather inconsequential; however, it should be noted that a single B-2 bomber costs approximately $1 billion, and the cost of a single major civil aviation accident can exceed $500 million in total financial losses. The costs in terms of personal suffering are often inestimable (Lauber and Kayten, 1988). In addition, the impact of a catastrophic mishap on the public's support of military aviation or on the revenues of an airline is probably severe. Although no concrete figures are available, it is likely that substantial public-relations "fallout" resulted from events such as the crash of Korea Air flight 801 in which 228 people died (NTSB, 2000); the near crash of China Airlines flight 006 in which two people were severely injured and other passengers were traumatized (Kolstad, 1989); or the accident involving Colgan Air Flight 3407 in which 50 people were killed (NTSB, 2009). In each of these cases, crew fatigue from long duty periods and/or circadian factors were implicated.

Extended Work and Insufficient Sleep are Common in Aviation

Extended duty times (work shifts that exceed eight hours) are already common in aviation, and increased demands on both civilian and military pilots will no doubt require additional work hours in the future. In civilian aviation, Gander et al. (1998a) found that one sample of pilots involved in short-haul trips worked an average of 10.6 hours per day, while another sample of long-haul pilots worked an average of 9.8 hours per day (Gander et al., 1998b). Rosekind et al. (1994) indicated the duty times of long-haul pilots in their cockpit napping study ranged from 8.4 to 14.8 hours per day. No doubt, many of these pilots were continuously awake for several more hours beyond the 11–15 hours logged as "duty time" considering that commute times and other nonwork activities are not considered "duty." Gander et al. (1998b) found that the average period of wakefulness for her subjects was in excess of 20 hours per duty day. Similar statistics for military personnel are not readily available, but it is a certainty that during times of military conflict, the situation is worse than what has been observed in civil operations. This no doubt accounts for the fact that a survey of Army pilots revealed that 81 percent thought fatigue was a contributing factor to increases in aviation accidents/incidents, 73 percent felt there was a widespread problem with fatigue in the military aviation community, and 61 percent were concerned that their own safety had at some point been compromised by fatigue or the lack of adequate rest (Caldwell et al., 2002).

Lengthy Duty Periods Have Been Associated with Performance Declines

The effects of extended work schedules are not fully understood at present; however, Rosa and Bonnet (1993) found that prolonged work shifts (greater than eight hours) led to decrements in alertness and performance in an industrial setting. Akerstedt (1995b) points out that long work hours may be associated with increased sleepiness, and Morisseau and Persensky (1994) found that overtime in the nuclear industry is related to an increase in incidents. Although such findings are at odds with those of Harrington (1994) who concluded that 10–12 hour work shifts were not associated with increased health or safety risks, Hamelin (1987) demonstrated a relationship between longer work hours and an increased risk of truck accidents, particularly at night. Samel, Wegmann, and Vejvoda (1997) have shown that pilot fatigue increases progressively as a function of flight length, and Rosekind et al. (1994) revealed that some pilots experienced increased performance lapses during the latter portion of long-haul flights. After examining Part 121 aviation human-factors accidents that occurred in the US between 1978 and 1999, Goode (2003) concluded the data "point to increased risk of accidents with increased duty time and cumulative duty time." (p. 312).

Time of Day is an Important Determinant of Fatigue and Performance

Working at times that are incompatible with circadian rhythms can produce problems that are separate from those associated with simply being awake or on the job for a long period of time. Night-shift performance is generally poorer than daytime performance regardless of the nature of the work. Primarily because of increased sleepiness at night, the probability of accidents on the highways, in industry and in aviation is increased (Akerstedt, 1995a). Monk and Folkard's (1985) characterization of the impact of nighttime work on even the most simple tasks raises serious concerns for pilots and crewmembers responsible for sophisticated modern aircraft. If the speed of answering a telephone switchboard is reduced, the frequency of meter-reading errors is increased, the speed of spinning thread is decreased, the ability to stay awake while driving is hampered, and the vigilance of train drivers is compromised at night, it is easy to imagine the potential impact of decreased nighttime alertness on a flight crew's ability to land a multi-engine jet carrying 200 passengers, fly a helicopter firing guns and missiles while hovering 50 feet above the terrain, or piloting a long-range bomber mission to drop weapons on a distant target. Dinges (1995) has shown that non-traditional work hours in combination with increased automation have substantially increased the risk of fatigue-related problems throughout the industrialized world. Furthermore, there is evidence that a number of high-profile catastrophies (that is, the grounding of the Exxon Valdez, the space shuttle Challenger accident, the crash of Korea Air flight 801, and the near meltdown at Three Mile Island) were at least partially attributable to the fatigue associated with night work (Mitler et al., 1988; NTSB, 1990; NTSB, 2000).

There is considerable evidence that night flights are especially vulnerable to cognitive lapses or "microsleeps" – brief periods during which sleep uncontrollably intrudes into wakefulness. Moore-Ede (1993) found that while microsleeps occurred in the simulator cockpit regardless of the time of day, there was a tenfold increase between the hours of 0400 and 0600. During this time, pilots made the greatest number of errors. Klein, Bruner, and Holtman (1970) reported that pilots' abilities to fly a simulator at night decreased to a level comparable to that observed with a blood alcohol content (BAC) of 0.05 percent. Wright and McGown (2001) found that while the sleepiness of long-haul pilots increased during both daytime and overnight flights, the occurrence of sleep was more frequent on flights that departed late in the night compared to those that departed earlier. Disturbingly, many of the microsleeps were so short (less than 20 seconds) that the crewmembers may not have been aware that they had briefly fallen asleep. Samel et al. (1997) reported evidence of inadvertent sleep in pilots flying nine to ten-hour night flights. Similarly, Rosekind et al. (1994) found a substantial increase in micro-events (slow-wave brain activity and/or slow eye movements) on long-haul flights, with night flights being particularly affected compared to day flights. Vigilance performance and subjective alertness ratings were degraded more at night as well. Caldwell, Hall, and Erickson (2002) found that the combination of sleep loss and night flying significantly accentuated the type of slow-wave electroencephalographic (EEG) activity that has been associated with insufficient alertness while concurrently causing the types of mood and cognitive deteriorations that impair crew coordination and responses to system deviations or failures. According to a report from the European Cockpit Association (2012), surveys among pilots reveal that night flights or a series of night flights are considered to be among the major contributors to fatigue.

Aircrew Fatigue: The Bottom Line

Data published in the open scientific literature make it clear that aircrews are susceptible to fatigue because of a variety of factors including extended duty periods (with extended periods of wakefulness), circadian disruptions from rotating work/rest schedules and traveling across time zones, sleep restrictions associated with short layovers (Gander et al., 1998b) and/or military sustained operations (Krueger, 1989). These factors are often exacerbated by cockpits that are highly automated and sometimes cramped, poorly ventilated, noisy, and dimly lit (Battelle, 1998).

Unfortunately, the problem of aviator fatigue is complex. Economic pressures continue to force civilian air carriers to cut the numbers of flight-crew personnel while attempting to maintain or enhance existing levels of service. Competition among air carriers is fierce, and the public expects the same level of service despite changes in global economic factors. Meanwhile, government budgetary constraints continue to adversely impact the force structure and manpower of every branch of the military service while work demands continue to escalate. The Air Force cut 20,000 airmen in 2014, the Army shed over 60,000 soldiers between 2010 and 2014, and the Marines expect to cut more than 25,000 personnel by 2017 (Larter, 2014). Needless to say, aviation units are suffering as much as ground-based troops. In 2013, the US Congress allowed "sequestration" of Department of Defense funding to take effect, and the full impact of this action on military budget reductions has raised serious concern among senior leaders (Tilghman, 2014). Meanwhile, the high operational tempo persists, and the combination of more work and fewer resources has raised concerns about force readiness. Some worry that even the most elite warriors are being worn down by repeated deployments.

Across the military and civilian spectrum, people are simply working longer hours and attempting to accomplish a greater number of tasks than ever before in order to do their jobs under conditions that are less than optimal. This no doubt is at least partially responsible for reports by the National Sleep Foundation (NSF) that the adult US population has added 158 hours a year to the average working/commuting time since 1969. The NSF further suggests that Americans have reduced their average amount of nightly sleep by 20 percent over the past century in part because of the increased prevalence of night work or shift work (25 million US adults now routinely engage in shift work), and in part because of widespread efforts to "accomplish more with less."

Unfortunately, there have been costs associated with the hectic tempo of modern life. Half of the adult public reportedly experiences at least one symptom of insomnia at least a few nights a week, and 93 percent feel that not getting enough sleep impairs work performance (National Sleep Foundation, 2002). Thirty-five percent of surveyed Americans complain their sleep is "poor" or "only fair" (National Sleep Foundation, 2014), and almost half say they feel unrefreshed upon awakening (National Sleep Foundation, 2008). Furthermore, fatigue-induced problems have become a rapidly growing concern in all walks of life – on the highways, in industry, and in air and ground transportation.

What Can Be Done?

Are there ways to deal with the fatigue that has become so commonplace in the industrialized world? The simple answer is yes, but the implementation of available solutions is not straightforward. While the reasons for this are complex, it is clear that there are substantial societal and political barriers to effective fatigue management. One example from the aviation arena is the issue of cockpit napping, an effective counter-fatigue strategy that was proposed and validated by a world-renowned US scientist (Rosekind et al., 1994) many years ago. Numerous studies, many of which were conducted by US government researchers, have clearly established the efficacy of napping for sustaining and restoring the performance of fatigued personnel in various environments including the cockpits of passenger jets; however, disagreements about the implementation of such a strategy in US air operations has resulted in the lack of Federal Aviation Administration (FAA) approval despite the approvals of foreign regulatory bodies. The implementation of napping in other settings has proven difficult as well because of societal attitudes and workplace regulations. People who admit they need extra sleep are generally viewed as weak and/or lazy, and the majority of firms, to include the US government, enforce regulations that require the immediate termination of anyone found sleeping in the workplace. Even many sleep disorders centers and sleep laboratories prohibit napping at work in spite of the fact that they recommend napping for some of their patients, and despite their admonishments that corporations should establish nap facilities and napping opportunities for shift workers.

Clearly, there is need for the scientific, medical, and industrial communities to reach a consensus about the problem of fatigue and what can be done to solve it. As is evidenced by the growing emphasis on the development and implementation of data-driven, scientifically valid, and collaborative fatigue risk management systems for aviation and other sectors, there is a growing concordance with regard to the importance of recognizing and managing fatigue in safety-sensitive contexts; however, in general, progress remains rather slow. Not only have corporate, union, and military leaders been unable to agree on the best remedies for this ubiquitous phenomenon, but the workforce itself (and the general public) has been slow to insist that something be done while at the same time rejecting cultural stereotypes that glorify and reinforce a macho attitude toward dangerous sleeplessness. The root of the problem is that the hard-charging, success-oriented people who make up the modern industrialized community and the world's military forces have yet to fully recognize the threat that human fatigue poses in terms of safety, health, efficiency, and productivity. In addition, society has been slow to accept the fact that fatigue stems from physiological factors that cannot be negated by willpower, financial incentives, or other motivators. Furthermore, members of the engineering community have long promoted the idea that many of the problems stemming from human error will ultimately be overcome by technological innovation (that is, engineering the man out of the system), and they have convinced a number of both civilian and military leaders that this assertion is true. This type of thinking first overestimates our ability to create and refine intelligent automated systems, and second, prevents the development and implementation of feasible and effective alertness-management strategies for the human operators who are responsible for flight (and industrial) safety now, and who will remain responsible for decades to come.

The next chapters will generally discuss the topic of fatigue/alertness management in non-technical language, explain why we should be concerned about the problem of fatigue in aviation and other settings, identify some of the primary factors that cause fatigue in the workplace, and offer information on fatigue countermeasures that can be implemented in ground-based and in-flight aerospace operations. The majority of the information is designed to be relevant for both civil and military aviation settings, but there are occasionally portions that will be more appropriate for one community than the other. At the outset, it should be noted that while technical jargon has been intentionally kept to a minimum and scientific references have been omitted from most of the remaining text in this book, the information contained in these chapters is based on scientifically established facts and not simply the opinions of the authors or others. Readers who desire to delve further into any topic that may hold particular interest should consult the reference/suggested reading list that appears at the end.

One final note before tackling the issue at hand is a reminder that the world is made up of a variety of people, all of whom come from widely divergent backgrounds, work in hundreds of unique settings, and bring to the workplace their own individual gifts and limitations. Because of this, it is unlikely that any one compilation of facts, set of rules, or collection of recommendations will apply equally to everyone. So, with this in mind, read ahead to gain a basic appreciation of this problem we call fatigue, and pick and choose the solutions that will likely work best to ensure your success and the success of those who work for you or with you. As with everything in life, some trial-and-error may be required before the best solution is found, but with the right facts at hand, this process will be relatively brief and painless, especially in comparison to the benefits that will result in the long run.

Top Ten Points About the Introduction of Fatigue

- Recent mishaps have increased the focus on pilot fatigue.
- Societal, economic, and technological factors have exacerbated the problem.
- Military sustained operations and global deployments further add to the equation.
- Fatigue contributes to 4–7 percent of civil aviation accidents.
- Fatigue is cited in a number of Army, Navy, and Air Force mishaps.
- Extended duty periods and insufficient sleep are at the root of the problem.
- Body-clock disruptions from shift work and jet lag are contributing factors.
- Personnel cutbacks in the commercial and military sectors are not helping the situation.
- Technology cannot solve the fatigue problems and may even make them worse.
- The only solution is a scientifically validated alertness-management program.

2 An Overview of Fatigue

The problem of fatigue in aviation must be proactively confronted if we are to ensure maximum safety and efficiency, as well as optimum personal well-being. It is not enough to take care of the aircraft unless the pilots are receiving the attention they deserve. Routine medical exams guard against the possibility of performance-impairing disease, and random drug screens minimize the chances that ingestion of illegal substances will adversely impact reaction time and judgment. Even mandatory retirement is enforced to guard against the problems that some feel are likely to occur with advanced age. But is everything possible being done to reduce the threat of fatigue-related decrements? Are effective alertness-management strategies being employed to maximize on-the-job performance? What exactly is fatigue, how do we know it is present, and how much should we worry about it?

Fatigue Defined

At the outset, it is important to define what we mean when using the term "fatigue." This seems like a fairly straightforward objective, but a trip to the local bookstore or library quickly reveals a wide variety of definitions that may apply. So, arriving at a precise characterization of fatigue is not as easy as it would at first appear. In fact, the premier sleep expert in the world, Dr William Dement, has said he has trouble fully defining fatigue, and he has been researching the subject for 50 years! However, Dr Dement goes on to say that he believes 95 percent of what is called fatigue results either from sleep deprivation or undiagnosed, untreated sleep disorders rather than boredom, monotony, stress, or unclearly defined biological processes.

The International Civil Aviation Organization (ICAO) defines fatigue as: "A physiological state of reduced mental or physical performance capability resulting from sleep loss or extended wakefulness, circadian phase, or workload (mental and/or physical activity) that can impair a crew member's alertness and ability to safely operate an aircraft or perform safety-related duties." Fatigue is a major human-factors hazard because it affects most aspects of a crewmember's ability to safely perform his or her job.

Note that the ICAO definition includes both mental and physical fatigue. However, in the present context, we will make sharp distinctions between these two and focus only on the mental fatigue that results from sleep loss and circadian factors. Despite the fact that physical and mental fatigue can exist together, they are quite different from one another. If a person becomes physically exhausted due to high-intensity physical effort, he may no longer be able to perform physically demanding tasks, but his alertness and concentration nevertheless will likely remain intact. Although there is a lack of scientific consensus about the impact of intense physical activity on cognitive status, several authors have concluded that physical exercise often has little or no impact on mental performance. However, when someone becomes mentally exhausted due to sleep deprivation and/or circadian desynchronization, cognition and alertness will definitely suffer extensively while most aspects critical for physical performance will likely be preserved. Studies have overwhelmingly shown that sleep loss degrades mood, mental capabilities, memory, and alertness; but 30 to 72 hours of continuous wakefulness does not reliably affect cardiovascular and respiratory responses, aerobic and anaerobic performance capability, muscle strength, or electromechanical responses.

At any rate, almost every pilot routinely contends with long work periods, unpredictable schedules, and non-standard duty hours (that is, shift work), while cognitive overload and physical exhaustion are less frequently encountered in the modern cockpit. As for mental boredom, there is little doubt that this can be a serious problem in long-haul commercial and military flights, but it is pretty clear that boredom doesn't actually produce drowsiness or dozing off in people who are not sleep deprived to start with. So, we will focus on the fatigue that stems primarily from the schedules that support the aviation mission.

Quantifying Fatigue

The tiredness that affects pilot alertness and performance results from internal physiological changes that are not fully understood. However, it has been well established that there are manifestations of these changes that are well beyond the control of individual willpower, professionalism, or motivation. For instance, studies have demonstrated that fatigue is associated with slower brain activity, changes in eye movements and pupillary responses, and other physiological differences that persist despite the presence of worker incentives or considerable individual effort. Obviously, fatigue is more than just a state of mind. Unfortunately however, quantifying fatigue is difficult because no biochemical markers for fatigue have been discovered. There is no Breathalyzer™ for fatigue like the one that police officers use to identify alcohol-impaired drivers, and because of this, it is extremely difficult to precisely identify people who are too tired to fly or too fatigued to engage in other forms of mentally demanding tasks.

We often must rely on subjective impressions about our own abilities to perform, while civilian supervisors or military commanders must rely on inferences drawn from knowledge of prior duty schedules and work hours to decide whether the next flight can be safely accomplished. Everyone knows what it feels like to be fatigued, and sometimes we can detect the appearance of fatigue in someone else, but how can we know when the amount of fatigue has crossed the line from being simply an unpleasant feeling to being a hazard to safe flight?

The absence of objectively established "fatigue markers" has significantly complicated attempts to manage fatigue on the flight deck and elsewhere. In addition, it has hindered attempts to understand the full impact of tiredness on safety. Furthermore, the lack of a "blood test" for the presence of excessive fatigue has made it difficult to convey the extent of the problem to the general public, operators, and decision makers. Many people continue to feel that fatigue is not a genuine concern because the fatigue levels of those responsible for accidents cannot be measured through blood or tissue samples. Because of this, we cannot say "Aha! This guy scores a nine on the ten-point fatigue scale, so he must have crashed because he was too tired." However, despite the fact that we cannot determine the presence of fatigue through a blood test, we can be fairly certain about fatigue's contribution to a mishap by examining factors that suggest the person's level of impairment at the time. The Massachusetts company Circadian® suggests considering the following:

- consecutive hours of duty at the time of the accident;
- hours of duty during the days leading up to the accident;
- irregularity of the work/sleep schedule;
- time of day relative to the body's clock;
- amount of wakefulness since the last sleep period;
- amount of sleep that was obtained during the last sleep period;
- existence of a cumulative sleep debt prior to the accident;
- the degree of external stimulation offered by the job or work environment;
- the level of physical or mental stress preceding the mishap;
- environmental factors such as light and noise levels;
- drugs that were taken before the accident.

Note how many of these indicators suggest that sleepiness was the root cause, or that something about the job setting likely made the effects of the person's sleep debt more pronounced. These factors will be discussed in detail later. However, at this point, it is important to consider that, until recently, such issues were rarely given the credit they deserved by accident investigators. This led to an underestimation of the importance of operator alertness for on-the-job safety. Thanks to the growing body of research on this issue, there is now substantial evidence that fatigue/sleepiness can be a significant problem and that overly tired personnel have been at least partially culpable in some of the most memorable disasters in recent history.

Figure 2.1 Where is the dividing line between "this guy looks a little tired" and "this guy should definitely not be in the cockpit?"

Evidence That Fatigue Played a Role in Some Noteworthy Catastrophes

Outside of the aviation arena, consider the near meltdown of reactor unit 2 at Three Mile Island nuclear power plant. This incident occurred when night-shift personnel failed to properly identify and mitigate a problem with the reactor's cooling system – a failure that nearly liquefied the uranium fuel rods, almost causing a meltdown, and precipitating the venting of radioactive gasses into the atmosphere. Years later in Russia, the actual meltdown at Chernobyl likewise stemmed in part from the actions (or inactions) of night-shift workers who were performing a non-standard test of whether the power generated from a "coasting" reactor could supply enough feedwater to provide adequate reactor cooling in the event of a shutdown. Needless to say, the test failed, a complete meltdown occurred, and the resulting accident was directly responsible for the deaths of 300 people, over $13 billion in economic disruption, and a substantial number of cancers and birth defects in the surrounding population. Another fatigue-related mishap was the grounding of the Exxon Valdez oil tanker in 1989. Because of initial media coverage of this event, most people believe the accident was caused by an intoxicated captain, but the captain was not at the helm during the fateful night when this oil tanker crashed into the Alaskan shoreline. Instead, the sleep-deprived third mate was piloting the ship without the benefit of help from his supervisors (who had been awake long hours overseeing the loading of cargo). While essentially "asleep on his feet," he missed clear warning signals, failed to remain in the shipping channel that was visibly marked with navigational buoys, and ran the ship aground, contaminating 1,400 miles of shoreline, and costing Exxon over $8 billion in fines and other expenses. The NTSB ruled this to be a fatigue-related accident.

There was no "fatigue breathalyzer" to scientifically establish the amount of tiredness present in the people who were involved in these three disasters, but there is plenty of evidence that sleepiness and fatigue were major contributors. All three of the events occurred in the early-morning hours when alertness is known to suffer as a result of extended wakefulness and circadian factors. Shift work was clearly a factor, and this brought into play scheduling irregularities and the likelihood that pre-mishap sleep quality and quantity was not optimal. In the case of the Exxon Valdez, the work environment probably offered little stimulation to help overcome the third mate's existing sleepiness – dim lighting and a lack of social interaction certainly did not help him stay awake. Although in all three of the disasters the fateful chain of events began with problems often not directly attributable to human fatigue, the fact is that obvious courses of action were overlooked, mistakes were made, and violations of safe and standard operating procedures were committed – just the sort of behavior that should be anticipated in fatigued operators.

Typical Effects of Fatigue

Fatigue is known to degrade many aspects of mental abilities and performance as well as psychological well-being. Consider this list of known fatigue effects:

- accuracy and timing degrade;
- lower standards of performance unconsciously become acceptable;
- attentional resources are difficult to effectively divide;
- the ability to efficiently integrate information is lost;
- activities become more difficult to perform;
- performance becomes increasingly inconsistent;
- social interactions decline;
- attitude and mood deteriorate;
- the ability to logically reason is impaired;
- the ability to maintain a clear picture of the overall situation diminishes;
- attention wanes;
- involuntary lapses into sleep begin to occur.

At first glance, some of these problems may seem inconsequential, but a closer look reveals the potential severity of each, especially in flight operations. Degradations in accuracy and timing can lead to the entry of improper navigational coordinates or other information into on-board computers or can produce potentially disastrous delays in responses to in-flight emergencies. Of greatest concern are the outright pilot errors such as landing on the wrong runway or flying at incorrect altitudes. The acceptance of sloppy performance can lead to flight-path deviations that can have serious consequences in congested airspace. The loss of the ability to effectively divide and allocate cognitive resources ultimately diminishes the capacity for multi-tasking which leads to channelized attention. Pilots are then at risk of becoming preoccupied with one task, such as navigation planning or radio communication, to the exclusion of some other more important jobs such as actually flying the plane. Impairments in the capacity to integrate incoming information can result in the loss of spatial orientation and/or situational awareness. This is particularly a concern in the rotary-wing environment where flights are conducted at lower altitudes in closer proximity to obstacles such as hilly terrain, trees, buildings, and communications towers. Perceptions that routine activities are more taxing than usual can cause crews to shed important tasks because they seem like too much trouble to manage. Increased performance variability means that someone will appear to be alert and accurate one moment while deviating far beyond acceptable parameters the next (and it is impossible to estimate how long each of these phases will last). A lack of interest in social interactions and a poor attitude can lead to a breakdown of crew coordination. The inability to think clearly and logically can dangerously retard decision-making speed, lead to inaccurate conclusions, and result in unsafe actions or a complete failure to respond to rapidly changing circumstances. Insufficient attention can lead to procedural violations and ultimately to the lack of awareness that can allow easily correctable problems to develop into full-fledged in-flight emergencies.

One of the most dangerous aspects of operator fatigue is the increased tendency to unconsciously lapse into brief episodes of sleep in the middle of tasks that require a high degree of vigilance. Research has shown that microsleeps (brief periods of inadvertent sleep) increase in frequency as a function of hours awake and low points in the body's daily cycle. Tired people often spontaneously drift off to sleep for several seconds at a time even when they should be concentrating on the task at hand. Since many of these microsleeps are only a few seconds long, people may mistakenly assume that they are not very important; however, to gain an appreciation for the dangers of even short lapses, consider that when driving down the highway at 55 mph, a car can travel the length of an entire football field in only five seconds. That is a long way to go with the eyes closed while fast asleep, especially considering that oncoming traffic, bridge rails, telephone poles, and other hazards are only a few feet away! To bring the situation closer to aviation, just imagine how far an aircraft flying 160 knots on approach to landing will travel in only a few seconds. And before you convince yourself that no one would ever fall asleep on approach, realize that published studies have documented microsleeps in pilots between top of descent and landing despite the fact that there were passengers on board and research personnel were monitoring everyone in the cockpit at the time. Furthermore, frank episodes of microsleeps (lasting more than 20 seconds) have been observed in helicopter pilots flying only 2,000 feet above the ground, and occurrences of outright sleep (lasting several minutes) have been observed in the entire cadre of flight-deck personnel of large airliners only minutes after departure.

Fatigue-related/Sleepiness-related Problems are Extensive in Aviation

In general, chronic sleepiness has been identified as a major problem in both the military and civilian world, and the issue spans all walks of life from personal well-being, to workplace and highway safety, and ultimately to aviation. And fatigue-related problems are not restricted just to the extreme situations in which repeated episodes of chronic sleep restriction and constant exposure to rapidly rotating work/rest schedules are found. Instead, the evidence is mounting that even one period of continuous wakefulness of a day or less can compromise the most basic aspects of human performance. In light of the fact that fatigued people are often no better off than people who are intoxicated, it is little wonder that the NTSB has long identified fatigue as a probable cause or contributing factor in accidents across all modes of transportation. In fact, the NTSB has issued over 200 safety recommendations focused on fatigue in transportation, addressing hours of service regulations, scheduling policies, education and training, diagnosis and treatment of sleep disorders, research, and vehicle technologies. However, despite years of extensive concern over fatigue-related safety issues, the FAA in the US only recently updated hours of service rules for pilots and the Federal Motor Carrier Administration just issued new rules for commercial truck drivers. Meanwhile, a 2012 poll by the NSF revealed that 41 percent of pilots and roughly one-third of train operators were getting less sleep than they felt was adequate on workdays, and 23 percent of pilots said sleepiness impacted their job performance at least once a week.

Fatigue in Commercial Aviation

In aviation, fatigue has long been recognized as a problem by aircrews, but the impact of inadequate pilot alertness on safety historically has been underappreciated. In fact, 1993 was the first time that pilot fatigue was officially ruled to be a contributing factor to an aviation mishap. After the crew of American International Airways Flight 808, who had been on duty for nine hours (and was scheduled for five more hours), crashed their DC-8 on approach to landing in Guantanamo Bay, Cuba, investigators concluded that fatigue was a causal factor. Since then, other noteworthy mishaps have been chalked up at least partially to aircrew fatigue. For instance, fatigue was blamed at least in part for the 2009 crash of a Continental Connection flight (Colgan Air Flight 3407) in which 50 people were killed. This tragedy occurred as a result of the captain's disregard for established stall recovery procedures in favor of ignoring the stick-shaker warnings and subsequently overriding the activated stick pusher in an attempt to maintain a nose-up attitude despite a dangerous deterioration in airspeed on approach. Shortly thereafter the plane struggled to remain aloft for several seconds with the crew making no emergency declaration before crashing and exploding about five miles from the end of the destination runway. Prior to this incident, the crew of Corporate Airlines flight 5966 displayed a similarly alarming disregard for established flight procedures when they unwittingly crashed their Jetstream 32 into a line of trees after descending below glide path on approach. As a result, 13 people died. The 1997 crash of Korean Air flight 801 also was due in part to fatigue. In this case, the recovered cockpit voice recorder revealed that the captain was "really ... sleepy," and this likely contributed to his mistaken belief that the inoperable ILS glideslope for the approach runway was in fact in service. As a result, he made an incorrect and steep approach into Guam, which tragically ended in a crash into Nimitz Hill. Of the 254 people on board, 228 souls perished. Similarly, the 1999 mishap involving American Airlines Flight 1420 in which 11 people were killed was fatigue-related. In this case, the captain and first officer who had been on duty for more than 13 hours prior to departing Dallas–Fort Worth in route to Little Rock, Arkansas, attempted to land their MD-82 in unacceptably severe weather after also failing to deploy the auto-braking system. Consequently, the aircraft skidded off the end of the runway and crashed into the approach lighting system before finally coming to a stop on the banks of the Arkansas river.

Another highly publicized aviation mishap has by some sleep/fatigue experts been attributed to the effects of fatigue associated with circadian factors. The 1985 near crash of a China Airlines Boeing 747 (Flight 006) in which several passengers were injured or traumatized by a 32,000-foot loss of altitude over the Pacific Ocean, is a classic example of what can happen when skill and judgment are required at the wrong point in the body's daily rhythm. Although the NTSB ruled out fatigue as a contributor to this near disaster, the investigative narrative points out that the pilot in command was attempting to diagnose problems on the flight deck during his circadian low point. Interestingly, as the official investigators of this incident dismissed any possible role of fatigue in the pilot's often inappropriate actions, they concurrently expressed wonderment at why a highly experienced pilot and crew could manage to misdiagnose the reasons behind a progressive loss of airspeed and altitude, disregard the indications of perfectly functioning cockpit instrumentation, and ultimately place a 747 in such unusual attitudes that the safety of everyone on board was threatened and the airframe suffered extensive structural damage. In all probability, if this incident

Figure 2.2 The crash of Korean Air flight 801 was, in part, due to pilot fatigue
Source: National Transportation Safety Board (Public Domain).

were evaluated today, a large part of culpability would be placed on aircrew fatigue, rather than blaming the problems on the pilot's "over-reliance on flight-deck automation."

Fatigue in Military Operations

Are there examples from the military community as well? The answer is of course "yes," but unfortunately the narratives of military mishaps are often not freely available. However, fatigue is clearly a concern for the military. In the US Navy and Marine Corps for example, fatigue was identified as the leading cause of Class A mishaps (that is, damage exceeds one million dollars and involves fatality or permanent disability or destroyed aircraft) – among the mishaps that were associated with aeromedical factors – between 1990 and 2011. Concerns about the adverse impact of aircrew fatigue have long been recognized in military aviators because of the nature of military and wartime operations. Military pilots are often under significant pressure to fly long missions, day in and day out, under some of the most austere circumstances. Army pilots in particular often do not have the benefit of a fixed-base facility where there are buildings suitable for the establishment of sleeping quarters because Army aircraft, most of which are rotary wing, do not require runways. However, carrier-based Navy pilots routinely suffer sleep disruptions from heat, vibration, and catapult noise, and thanks to air-to-air refueling and long-range flight capabilities, Air Force pilots are regularly involved in lengthy missions, strapped into ejection seats with no sleep opportunities for hours on end.

Despite differences in aircraft and the facilities required to support them, one thing that all of the services have in common is that the pilots frequently have little choice except to make the flight at hand because their failure to do so can have consequences that extend well beyond the immediate safety of the flight crew or the economic concerns that place pressure on their commercial counterparts. If an airline captain declines a flight, the company and passengers likely will be upset and the flightcrew will no doubt have a lot of explaining to do. However, from a safety standpoint, the fact that an overly tired commercial aircrew has chosen to err on the side of caution can only save lives rather than place anyone at risk. In contrast, when a military pilot declines a mission, it is entirely possible that an enemy threat will not be removed, innocent people will not be defended, and wounded or injured troops or civilians will not be effectively evacuated to a safe location. Thus, the military aviator faces a very different cost/benefit equation than his civilian equivalent. In addition, the military aviator's situation is complicated by the fact that intense around-the-clock air operations are relied upon to stress enemy forces as well as to provide support for friendly troops operating 24 hours a day on the ground. Not only must the military pilot fly and fight at night (many times after a significant period of daytime duty or an overseas deployment), but this pilot must train at night as well. Due to these factors, it is understandable that fatigue has been identified as a contributor to safety problems in the Air Force, the Navy, and the Army. Unfortunately, all three US military services report fatigue-related mishaps that meet or exceed those of the commercial world.

Pilot Perceptions about Fatigue

In light of these data, fatigue concerns are justifiable based on accident statistics alone. But what do the pilots think about the extent of the problem? Surveys of airline transport pilots on this topic are quite limited; however, corporate pilots and some military pilots have been asked about the extent of fatigue-related concerns in their lives. A 2000 NASA survey of almost 1,500 corporate/executive aviators revealed that 74 percent considered fatigue a moderate or serious concern, 61 percent said fatigue was a common problem, and 85 percent felt fatigue was a moderate or serious safety issue. Furthermore, almost three-quarters of the pilots admitted that at some time they had been so tired that they had actually nodded off in flight. When asked to identify the causes of this on-the-job sleepiness, they pinned the blame on scheduling issues such as late-night arrivals and early-morning departures, operational factors such as unpredictable weather and excessive workload, and simply the fact that they were routinely unable to obtain sufficient sleep. A survey published in 2006 revealed that 75 percent of short-haul pilots complained of severe fatigue and 81 percent said their fatigue was worse at the time of the survey than it was two years earlier. A significant fatigue factor for respondents was the extent to which they were flying into their discretionary hours. In general, such results are consistent with those cited in the 2012 Barometer on Pilot Fatigue report, which documented surveys of pilots carried out by member associations of the European Cockpit Association. The executive summary of this report noted that over 50 percent of pilots overall felt fatigue impaired their flight performance; and that 43 percent of UK pilots, 50 percent of Norwegian pilots, and 54 percent of Swedish pilots report having fallen asleep involuntarily in the cockpit while flying. Sixty-five percent of French and Dutch pilots said they had experienced "heavy eyelids" in flight. Night flights or the requirement to make consecutive night flights were thought to be major contributors to fatigue reported by the respondents.

These problems are not limited to civil aviation. Similar results have been obtained from US Army aviators and crewmembers. Seventy-two percent reported having flown when they were so drowsy that they could have easily fallen asleep, 73 percent considered fatigue a widespread problem, and 61 percent expressed concern that their personal safety had at some point been compromised by fatigue. Disturbingly, almost half of these crews indicated that they had periodically fallen asleep while in the cockpit – less than the amount reported by the corporate/executive sample, but still enough to make any passenger, other aviator, or anyone on the ground more than a little nervous. Why are these military aircrews experiencing fatigue-related problems? The answers were similar to the ones offered by corporate/executive pilots. They blamed excessive fatigue on long work schedules, but added that improper sleep quarters (too noisy, too bright, or otherwise uncomfortable) were also problematic (Army pilots, for instance, often have to sleep in tents in the field, or as was the case in Operation Desert Storm, huge concrete parking lots where even the tiniest sound echoes throughout the whole place for two or three minutes at least). However, the bottom line for them was the same as for their civilian cohorts – the root cause is *insufficient sleep*.

Legal Accountability for Widespread Fatigue is on the Rise

Needless to say, fatigue is a safety issue for the aviation community whether military or civilian. This should be no surprise, since the reality is that fatigue is a growing concern throughout numerous sectors of the industrialized world. Beyond flight mishaps, there are plentiful sleepiness-related highway crashes, maritime and railway accidents, and industrial errors and injuries. However, in addition to safety hazards posed by tired people, the straightforward economic impact of fatigue has recently become a concern. According to the American College of Occupational and Environmental Medicine, fatigue issues (from all causes) are responsible for more than $136 billion per year in health-related lost productivity each year. This is nearly twice the estimate for workers without fatigue.

Eighty-four percent of these costs are due to lower performance while at work rather than absenteeism. Needless to say, lost productivity from any cause is a serious issue for the corporate world. In addition, a newer concern is the increased litigation aimed at fatigued operators who cause accidents, injuries, and deaths among coworkers or other innocent people. Society's long history of devaluing the importance of sleep while honoring those who work and stay awake long hours has in the past made the drowsiness from inadequate sleep a more acceptable excuse for falling asleep on the job, even if an accident resulted. However, over the past 20 years, there have been numerous examples of new societal attitudes that have led to the criminal prosecution of sleepy operators as well as monetary judgments against the perpetrators of fatigue-related mishaps. More to the point, the employers of people found to be at fault in fatigue-related accidents are beginning to face stiff fines for their contribution to the problem.

In 2014, Wal-Mart was sued for negligence after one of its truck drivers fell asleep and subsequently crashed into a celebrity's limousine, seriously injuring the celebrity and killing another passenger. The lawsuit requested punitive damages for what it referred to as Wal-Mart's "gross, reckless, willful, wanton, and intentional conduct" in having a driver on the job for over 13 hours.[1] It notes that Wal-Mart was aware that the driver had commuted 700 miles before beginning his shift and alleges that the company ignores the fact that workers routinely break established safety rules. In addition, a criminal complaint accused the driver of not sleeping for more than 24 hours prior to the crash in violation of New Jersey law. In a different incident, an Ohio hospital was sued by the husband of a nurse who was said to have been "worked to death" due to excessive overtime and extended shifts attributable to hospital understaffing.[2] According to a spokesperson for the union that represents nurses, "chronic understaffing is rampant throughout hospitals around the country," and in this case it is alleged that the hospital was aware that this particular nurse was being severely overworked, but failed to take action to address the situation. As a result, the nurse was killed when she fell asleep on the way home from work, veered off the highway, and crashed into a tree. The plaintiff (her husband) hopes that suing the hospital will focus attention on the fact that safe staffing ratios are largely unregulated in US hospitals.

These are just a few examples reflecting society's growing intolerance of fatigue-impaired operators who knowingly place themselves and others at risk by not properly attending to their own sleep needs and society's concerns over corporations who disregard the risks that fatigue workers pose to themselves or others.

Efforts are Underway to Address the Issue

The message that people need to be alert while performing their jobs and driving to and from work is beginning to penetrate the business world, the military, family life, and the lives of individual citizens. Thanks to organizations such as the National Sleep Foundation (NSF), society is learning that chronic sleepiness is associated with many costs – monetary as well as personal. Unfortunately, as was the case with drunk driving several years ago, it seems that it has taken serious accidents, incarcerations, and expensive fines to enforce the message that the NSF and sleep/fatigue experts have worked so hard to deliver.

Times are changing. A perfect and highly relevant example is the new set of scheduling regulatory requirements for part 121 air carriers and their pilots that was recently issued by the United States Federal Aviation Administration (FAA) in 14 CRF part 117, Flightcrew Member Duty and Rest Requirements. First, the new rule sets different requirements for flight time, duty period, and rest based on the time of day pilots begin their first flight, the number of scheduled flight segments, and the number of time zone transitions. Second, the length of the flight duty period now considers when the pilot's day starts, the number of flight segments that are planned, and whether the crew complement is augmented or not. The duty period takes into account the time before a flight or between flights without intervening rest whether the duty is normal operations, deadhead transportation, in-flight training, simulation training, or airport standby or reserve duty. Third, flight time is limited to eight or nine hours depending on the start time of the pilot's entire flight duty period, and a ten-hour minimum rest period (which must include an eight-hour opportunity for uninterrupted sleep) is required as well. Fourth, consideration for cumulative fatigue has been factored in by placing weekly, 28-day, and (in the case of actual flight time) annual limits on flight duty and flight time. There is also a requirement for at least 30 consecutive non-duty hours each week to allow sufficient recovery from cumulative fatigue. Finally, regulatory requirements have been established to ensure that both pilots and airlines share fitness-for-duty responsibilities, and there are provisions for airlines to implement Fatigue Risk Management System (FRMS)-based fatigue-management alternatives based on science in combination with validated data and continuous FAA monitoring for operations that exceed the limitations of the rule. All of this comes with a congressionally mandated requirement (Public Law 111–216, § 212(B)) requiring all part 121 air carriers to develop, implement, and routinely update its Fatigue Risk Management Plans (FRMP) that includes educational efforts aimed at addressing the causes and effects of fatigue along with developing and implementing fatigue countermeasures, including the impact of overwork, commuting, or other day-to-day activities that routinely occur within the structure of the regulation. The overall objective of the congressionally mandated FRMP is to increase pilot alertness in an effort to reduce pilot performance errors.

Similar changes to those being implemented in US aviation are taking place in other parts of the world, and they include not only transportation operations, but petrochemical, agricultural, and industrial organizations as well. Hopefully, these and other efforts are evidence that society is increasingly aware that sleepy people don't belong in the workplace, and that everyone must share the responsibility for maximizing alertness at home, on the highways, in the air, and everywhere else!

Part of the remaining problems may be that we have all become so habitually tired that we no longer recognize the extent of our own sleepiness. So, before finishing this chapter, perhaps some honest self-evaluation is in order. Take the Epworth Sleepiness Scale contained in Table 2.1, add up the responses to each item, and compare your results to the typical scores from normally-alert people and those who are suffering from sleep disorders.

Is your score normal (10 or less), or is it more like extremely sleepy people with clinically diagnosed sleep disorders of narcolepsy or idiopathic hypersomnia (17–18)? Maybe you're somewhere in between, with a score of 11–12. This would make you similar to sleep-apnea patients who suffer from severe daytime sleepiness because they wake up numerous times every night due to breathing problems. Or perhaps you are like people who have periodic limb movement disorder. The fact that these patients suffer from sleep that is chronically disrupted due to muscle twitches in their arms or legs usually makes them score in the 9–10 range. How sleepy is *too sleepy* to be safe? The bottom line is that a score of more than 10 means that day-to-day tiredness has reached problematic levels. If this is the case, something has got to change to reduce the possibility of a fatigue-related mishap, or if nothing else, to simply reduce the concentration difficulties, immune-system

Table 2.1 The Epworth Sleepiness Scale

How likely are you to doze off or fall asleep in the following situations, in contrast to feeling just tired? This refers to your usual way of life in recent times. Even if you have not done some of these things recently, try to work out how they would have affected you. Use the following scale to choose the *most appropriate number* for each situation:

0 = would *never* doze 1 = *slight* chance of dozing
2 = *moderate* chance of dozing 3 = *high* chance of dozing

Situation chance of dozing	
Sitting and reading	________
Watching TV	________
Sitting, inactive in a public place (e.g. a theater or a meeting)	________
As a passenger in a car for an hour without a break	________
Lying down to rest in the afternoon when circumstances permit	________
Sitting and talking to someone	________
Sitting quietly after a lunch without alcohol	________
In a car, while stopped for a few minutes in the traffic	________
Total score	________

Source: Reprinted with permission from Johns, M.W. (1991), 'A New Method for Measuring Daytime Sleepiness: The Epworth Sleepiness Scale,' *Sleep*, Vol. 14, pp. 540–545.

impairments, family discord, and/or social/interpersonal problems that tired, irritable, sleep-deprived people often suffer.

One of the first things that must change is the widely held belief that if you live with fatigue long enough, you will adjust, and it just won't be a problem any longer. As the next chapter will demonstrate, this is definitely not the case.

Top Ten Points About the Overview of Fatigue

- Ninety-five percent of fatigue results from inadequate amounts of sleep.
- There is no "blood test" for fatigue, but its impact can be estimated from situational indicators.
- Fatigue played a role in the Three Mile Island, Chernobyl, Exxon Valdez, and other famous accidents.
- Research proves that cognition, mood, crew coordination, and attention are compromised by fatigue.
- Tired pilots have been shown to involuntarily doze off in the cockpit.
- Fatigue has been implicated in mishaps/incidents such as Colgan Air Flight 3407, Korea Air Flight 801, American Airlines Flight 1420, and China Air Flight 006.
- On US highways, fatigue causes 100,000 crashes and 1,500 fatalities annually.
- Three-quarters of corporate/executive and 72 percent of Army pilots feel fatigue is a serious concern, and 85 percent of short-haul pilots complain of severe fatigue.
- An amazing number of civilian and military pilots admit they have fallen asleep at the controls.
- Lawsuits against fatigued operators and their employers are on the rise.

Notes

1 See http://www.reuters.com/article/us-people-tracy-morgan-idUSKCN0QG1HR20150811#pxYDVgViECQIGTlY.97 (accessed 12 December 2015).

2 See http://www.cnn.com/2013/11/12/health/ohio-nurse-worked-to-death-lawsuit-says/ (accessed 12 December 2015).

Part II

Causes of Fatigue

3 The Nature of Fatigue

What is all of this talk of being fatigued? Why does it seem that everyone is complaining about being so tired? Could it be that the fast pace of the modern world is just too much, or is that people are not as tough as they once were? Is it possible to solve our fatigue problem with the right kind of thinking, enough motivation, or a well-planned training program that teaches us how to work no matter how tired we are? Or are we really just fatigued because we are only human and we're trying to do too much?

Fatigue is More Than a State of Mind

There has long been the misconception that fatigue is simply a state of mind rather than something "real." Because of this, it should come as no surprise that leaders and a lot of individuals have the opinion that the right amount of professionalism, enough dedication, and plenty of motivation will overcome the problem of tiredness in the workplace. People who complain of too much work or too much overtime, shifts that are overly long, or too-frequent requirements for night duty are often seen as lazy malcontents who just don't have the "right stuff." And it isn't just employers, supervisors, or military commanders who sometimes think this way. In fact, a lot of individuals think they can adjust to overwork and sleep deprivation if they can just hang on long enough to get used to it. They believe that training to get by with less sleep is similar to training for an athletic event. No pain, no gain! They think it might be unpleasant to try to get by on less than their required daily amount of sleep at first, but after a while, the body and mind will get stronger and adjust. Then, those "wasted hours in bed" can be turned into productive work hours, quality family time, and/or relaxing recreational activities.

Fatigue is Based on Physiological Factors

Faulty views about the nature of fatigue were once pervasive, and many people still hold them today. They still believe they can learn to live with less sleep if they just will themselves to feel more alert or that they can sleep-deprive themselves enough to adapt. With thinking like this, it is no wonder we have a national sleep debt that is phenomenal! The truth of the matter is that the drive for sleep is not a product of imagination and it is not under voluntary control. Instead, it is a basic physiological reality just like the need to eat and drink. In order to survive and function effectively, living things must remain properly hydrated, they must consume adequate fuel in the form of food, and they must obtain a certain amount of sleep every day. If they neglect to do these things, performance, well-being, and even the chances of survival will suffer. The body signals when fluid levels are low by producing the sensation of thirst which prompts drinking behavior. The body strives to maintain a proper energy balance by generating the sensation of hunger when a caloric deficit is present. Likewise, the body signals the need to sleep by producing sensations of drowsiness, or in extreme cases, a virtual inability to remain alert enough to perform even the most basic tasks.

The Physiological Drive for Sleep is Similar to the Drive to Eat

Is it reasonable to think that chronically ignoring the symptoms of any type of physiological deficit is going to somehow change the underlying need that is responsible for these symptoms? The answer is "no." As a case in point, consider the physiological "set point" for body weight maintenance. This "set point" phenomenon can best be illustrated through an examination of the present-day problem of obesity in America. The source of the problem is pretty simple: Overweight people are consuming more calories than they are burning, and the surplus is being stored as fat. If overweight people are to lose weight, they just need to start taking in less than they are using. It is as simple as that! However, the problem is not with convincing people what they should do, but figuring out how to enable people to actually accomplish what they already know is the best course of action. The fact of the matter is, the brains of these overweight people strive diligently to prevent drastic changes in body weight. A great deal of research has been conducted to determine how to make people successful at reducing the amount they eat (and sticking with a new diet for the rest of their lives), but the greatest roadblock is that people become chronically and persistently hungry when they ignore what many scientists consider to be an underlying physiological "set point."

Back around the time of the Second World War, Ancel Keys performed ground-breaking research in the areas of hunger and nutrition, and this is what he found: When a group of volunteers received a 50 percent cut in available calories, they lost half of their body fat in about two months. However, their behavior and mood changed as well. As the study progressed, the participants became irritable, withdrawn, and sedentary, avoiding almost anything associated with energy expenditure. Meanwhile, food became the most important thing in their lives. They thought about it all the time, they were obsessed with references to food in books and movies, and some said they would cook as a recreational activity once the study was over! These symptoms are very familiar to anyone who has been on a diet. When the volunteers were again fed enough to start regaining the lost weight, it is interesting to note that their mood and hunger problems did not immediately subside despite the acute caloric surplus. Although they were consuming far more calories than they were expending, they remained ravenous until almost a year later when their body weight finally returned to normal. They never learned to ignore their insufficient food intake or the hunger that resulted. Their bodies never "adjusted" to being starved even though they were food deprived for a long time. In fact, they became psychologically obsessed with obtaining the very thing that they had been forced to live without! This is what happens when overweight people attempt to loose a significant number of pounds. If Ancel Keys had deprived people of water, they no doubt would have become obsessed with obtaining enough to drink. Or, if they had been deprived of sleep, his volunteers' bodies would have screamed for a night of restful slumber, and based on what we know today, his participants likely would have uncontrollably begun to doze off despite external penalties or heroic efforts to the contrary. How many of us long for the next weekend or the next vacation just so we can "catch up on sleep?" How many mornings are characterized by repeated attacks on the snooze button before enough willpower develops to drag ourselves out of bed? The simple fact is that no one gets used to not getting enough sleep. They might be able to do it, but they never overcome the drive for sleep or the consequences that invariably follow sleep restriction.

Everybody Has a Biological Sleep Requirement

Everyone is born with a certain sleep requirement that must be fulfilled in order for them to feel their best and function effectively on a day-to-day basis, year in and year out. In addition, after the adolescent years, this sleep requirement remains fairly constant throughout life, and there is nothing that can be done to change it. Furthermore, exposure to chronic sleep deprivation cannot teach people to get by on less sleep than they physiologically need. In fact, the longer someone goes without sleep, the sleepier they get. The misconception that people can learn to need less sleep or that they can overcome sleepiness through some type of adaptation procedure is just as dangerous as convincing people that they really do not have specific dietary requirements or that they can learn to live with a chronic state of dehydration. Sleep-deprived people are sleepy people, and sleepy people are not working or living up to their optimal potential, just like starving people are dull, listless, unmotivated, and cognitively and physically impaired.

Everyone's Alertness is Affected by the Body's Clock

Everyone also has an internal body clock that affects their alertness and activity differently throughout the 24-hour cycle. The misconception that circadian rhythms can be overcome by sheer willpower is another factor that has produced problems in the workplace. Adaptively shifting to a new daily sleep/work schedule, particularly a non-standard one, involves far more than just rearranging the daily routine. Traditionally, there has been a rather poor general understanding of the fact that humans are physiologically programmed to be alert at certain times of the daily cycle and drowsy (or asleep) at other times. In addition, leaders and employees have only recently begun to gain a better appreciation for the fact that it takes a while (often many days) to adjust to any new sleep/work schedule, and that most people never fully adapt to working at night and sleeping during the day.

Evolution Has Not Prepared Humans to Deal Well with Time Changes

One reason for our inability to quickly adjust to new schedules may be the fact that, relatively speaking, humans have only been attempting to conquer the issues surrounding shift work for a short time. Consider the fact that people were sleeping at night and working (or active) during the day for thousands of years before the 1883 invention of the electric light bulb literally changed things overnight. Before artificial light sources effectively turned night into day, people went to bed shortly after dark because they could not see well enough to do much else. Furthermore, they slept (or at least remained inactive) until the sunrise provided sufficient light to get back to work. While it was dark, factories could not produce products, business offices could not do paperwork (at least easily), and farmers could not farm, so everyone rested. However, when the electric light bulb was created, the seeds were planted for the development of the 24-hour society that now exists throughout the industrialized world. The light bulb enabled around-the-clock operations in almost every sector of society, and since people *could* work around the clock, they *did* work around the clock. In the military, a similar change in work dynamics occurred in the 1950s with the advent of night-vision technology. Before night-vision devices were available, military operations were either substantially curtailed or entirely suspended during the hours of darkness because it was difficult to fight an enemy that could not be seen. However, thanks to night-vision technology, night fighting has become a ubiquitous component of effective combat strategy.

Technology Has Had Both Positive and Negative Effects

The electric light, night-vision devices, and other technological advancements have been responsible for substantial progress in both the civilian and military worlds. So-called "24/7 operations" have increased overall productivity while creating greater efficiency. Extremely reliable machinery has enabled a continuous work flow without lengthy periods of maintenance "downtime." Rapid and reliable air travel has enabled us to cross numerous time zones in only a few hours. Night-fighting strategies on the battlefield have increased the mobility and lethality of military forces while improving our defensive posture. However, there have been costs associated with these enhanced capabilities. The ability to work around the clock has forced the people who support 24/7 operations to disregard the genetically programmed sleep/wake cycle in an effort to keep the factories running, the store shelves stocked, the airplanes flying, and the forces fighting. The ability to rapidly traverse time zones has outstripped our capacity to adjust to different day/night schedules. The result has been an increase in sleep problems, less-than-optimal mood and performance, and for some, a definite decrease in quality of life.

The requirement for people to work shifts that are in opposition to their biologically determined periods of alertness and drowsiness, and the requirement to switch time zones more rapidly than humans are able to adapt, has led to performance problems, psychological difficulties, and even decrements in immune-system functioning. Although it was once thought that the feelings of tiredness and discomfort associated with shift changes and transoceanic jet travel were simply a state of mind, science has proven that this is far from the truth. No matter how well personnel are compensated, how professional they are, or how much they desire to do their best, physiologically based limitations threaten to undermine their performance and safety. In fact, in 2010, the International Agency for Research on Cancer said shift work that disrupts the circadian cycle is a probable carcinogen!

Appreciation for the Underlying Factors is Essential to Addressing Pilot Fatigue

Why do we have these limitations? Considering the issue from an evolutionary standpoint, it makes intuitive sense that people would not easily adjust to 24/7 operations or rapid time zone transitions. For thousands of years, people worked during the day and slept during the night and there was consistent synchronization between the environmental rhythms and the body's rhythms. But in little more than 100 years, work hours and travel capabilities changed exponentially, while at the same time, basic human capabilities remained the same. People were not designed to operate around the clock anymore than they were equipped to suddenly shift their daily work/rest schedules to new time cues overnight. These limitations have created a host of challenges, but these can be met once an appreciation of the basic underpinnings of the problem is gained.

The next chapters will further define the underpinnings of the fatigue and circadian-rhythm disruptions that threaten to affect performance and quality of life. There are no easy solutions to the general problem of pilot fatigue because shift work and time zone transitions produce circadian disruptions while early-morning report times, long duty days, uncomfortable sleep environments, and a host of other factors combine to increase the drive for sleep. However, the probability that fatigue will build to dangerous levels can be effectively reduced when individuals understand the basics of fatigue and assume responsibility for maximizing their own sleep quality. Similarly, fatigue-related risks can be controlled when organizations make a concerted effort to avoid (or at least minimize) fatigue-producing practices in their day-to-day operations.

Forget About the Myths That Undermine Effective Alertness Management

Private lives that push the limits of human endurance for the sake of work and play at the expense of sleep and rest predispose individuals to poor health, psychological distress, social maladjustment, and risky job performance. Flight schedules that fail to consider human sleep needs and circadian rhythms create fatigue-related problems in the cockpit that place crews at risk for disaster (or at least some very close calls). On-board crew-rest facilities that are designed without attending to the relationship between environmental factors and restorative sleep may have immediate economic benefits, but will backfire in terms of employee dissatisfaction, poor crew readiness, and impaired safety. Arranging long-range military deployments without considering the impact of long duty hours and the problems associated with desynchronized body clocks will adversely impact operational readiness.

Meeting the transportation demands of the 24/7 society and supporting military missions to maintain peace and order in the world will no doubt continue to strain the capabilities of modern aviators. However, despite unavoidable operational constraints, an effort to resolve misunderstandings about human physiology and the basic nature of fatigue in aviation will reduce this strain to more manageable levels. Dr David Dinges, one of the best-known sleep and fatigue researchers, has identified six common myths that are often at the root of the problem. The effective elimination of these misunderstandings will go a long way toward maximizing the alertness, performance, and psychological well-being of pilots in all sectors of present-day aviation.

Myth 1

It is often assumed that if an airframe can be readied for service at a moment's notice, day or night, the pilot should be ready to go as well. However, the truth is that even the most highly skilled aviators are affected by the same biological limits that influence the alertness levels of everyone else. Simply stated, fatigue accumulates with increased hours of continuous wakefulness, and the body's natural rhythms make sleepiness more of a problem at some times of day than at others. The body cannot adjust quickly to changes in work hours, and forcing it to do so increases the risk of falling asleep at the controls or lapsing into sleep while performing other in-flight duties.

Myth 2

There is an assumption that good aircraft maintenance will ensure a high degree of operational safety. The reality is that even the most expertly maintained airplane or helicopter cannot function properly when it is being operated by an individual who is falling asleep at the controls. Errors of inattention and response failures are among the most common causes of serious accidents, and these types of problems have little to do with the state of the equipment being operated.

Myth 3

It has been suggested that a high degree of training, combined with past experience with sleep deprivation and shift work, is the key to avoiding performance problems associated with fatigue from overwork and rotating duty schedules. However, it is clear that people cannot be trained to overcome the effects of on-the-job sleepiness, despite familiarity with the problem and despite the fact that they may ultimately accept difficult work schedules as being "just a part of the job." It has been shown that sleep-deprived people accumulate a substantial sleep debt over time (cumulative sleep loss) that degrades their performance and increases risk by concurrently reducing their ability to accurately judge their own level of impairment.

Myth 4

Some people tend to believe that increased pay or rewards can overcome fatigue by increasing the motivation of sleepy crews. In fact, while it is true that extra incentives may in the short run encourage people to work harder to compensate for biologically driven losses of alertness, the beneficial effect is temporary at best. Overtime pay may increase someone's willingness to accept difficult assignments, but it has little effect on how safely these assignments are completed.

Myth 5

When establishing new work/rest schedules, there sometimes is an assumption that determining the amount of rest or sleep employees will need in order to avoid performance problems is a straightforward matter. In addition, there is a tendency to equate "time off" with sleep. In fact, *rest* is not equivalent to *sleep*, and there is no guarantee that the time off will be used to gain sufficient sleep to return to work alert and refreshed. Also, even if it were possible to ensure that aircrews would sleep a specific amount of time, it is quite difficult to determine exactly how much sleep would be necessary to restore the alertness of each individual. Some people may require only seven hours of sleep per day while others need nine. In addition, it makes a difference where in the 24-hour cycle the sleep period is placed.

Thus, because of the body's internal clock, a set amount of daytime sleep is unlikely to be equivalent to the same amount of sleep at night.

Myth 6

Some people believe that fatigue really should not be a concern since fatigued aircrews often safely complete trips without having accidents. Although it is true that sleepy operators may safely reach their destinations, there is little doubt that a loss of alertness increases the risk of problems associated with falling asleep at the controls, missing radio calls or navigational checkpoints, and reacting too slowly. Drunk drivers do not usually crash on their way home from the local bar; however, most of us probably would decline an invitation to ride in a vehicle driven by someone who was clearly intoxicated. Recent research has shown that there is reason to be equally concerned about fatigued pilots.

These myths must be dispelled at the management and operational levels of both civilian and military organizations in order to save lives and improve both safety and performance in the long run. Managers, commanders, and individuals who educate themselves and their peers regarding what is currently known about the nature and effects of fatigue, while supporting efforts to further enhance our understanding of fatigue-related problems, are key to the development and implementation of effective and feasible fatigue countermeasures.

Top Ten Points About the Nature of Fatigue

- Fatigue is a result of physiological factors and not just a "state of mind."
- People cannot train themselves to get by on less sleep by constantly sleep-depriving themselves.
- Fatigue has been exacerbated by technological advances that have outstripped the capabilities of humans to rapidly adapt.
- Humans were not designed to work at night or quickly transition to different sleep/wake schedules.
- Even under the best circumstances, nighttime alertness will be less than alertness during the day.
- It takes time for the body clock to adjust to any new work shift or any multi-time zone change.
- There is no 'magic bullet,' because biological factors underlie the degree of sleepiness/alertness.
- Professionalism and training cannot override the physiologically-driven sources of fatigue.
- Monetary and other incentives cannot overcome biologically-based pilot fatigue.
- Fatigue is a clear threat to air safety despite the fact that tired pilots don't always have accidents.

4 The Processes Underlying Sleepiness (Fatigue) and Alertness

As was generally alluded to in the preceding chapter, there are basic physiological factors that determine the level of alertness at any given time in the daily 24-hour cycle. Two of these factors are particularly relevant for a discussion of general sleep needs and shift work, but a third will be briefly included as well for reasons that will become apparent later on. When attempting to understand the basic nature of fatigue, the impact of each of these factors must be considered.

The Primary Factors That Affect Alertness/Sleepiness

The impact of each of these factors is very real, and everyone has to contend with their effects throughout life. The first factor, time since last sleep, is responsible for what has become known as the homeostatic drive (or mechanism), and the second factor, time of day according to the body clock, is referred to as the circadian drive (or mechanism). These drives, as defined above, pretty much determine how much sleepiness is present at any particular time of day. The third factor, sleep inertia, or the short-term grogginess that occurs immediately upon awakening from any sleep period, is not a concern in most workplaces; however, in situations where in-flight napping, in-flight bunk rest, or other types of on-site workplace naps are used as a fatigue countermeasure, the potential impact of sleep inertia should be considered. Besides straightforward factors such as the amount of wakefulness, the amount of sleep, and the amount of time from awakening until beginning job performance, there are influences such as the *quality* of off-duty sleep and the *continuousness* of work that can also impact the amount of drowsiness experienced during the hours of wakefulness, and these will be discussed in more detail later.

Table 4.1 The three primary factors that affect alertness/sleepiness

1. Homeostatic sleep pressure – the amount of time awake since the last sleep period
2. Circadian factor – time of day according to the body clock
3. Sleep inertia – the short-term grogginess right after awakening

Homeostatic Mechanism

The homeostatic drive is basically determined by the length of the period of continuous wakefulness. In other words, the drive for sleep tends to be relatively low immediately after awakening since the sleep need has just been fulfilled. However, as the day passes, the need for sleep gradually increases. If the period of continuous wakefulness extends beyond a typical 16-hour day, the homeostatic drive becomes particularly noticeable as the maintenance of wakefulness becomes more and more difficult. A good analogy for the homeostatic mechanism of sleep/wakefulness is the drive for hunger. Immediately after eating, the state of hunger tends to be pretty low, but as the time since the last meal increases, so does the drive to eat once again. Of course there are other factors that have an influence on the level of sleepiness or alertness at any given time of day just as there are other factors that influence our appetite for food. Some of the things that may make people feel sleepier than would be expected (or at least increase their awareness of their existing sleep debt) include how well they slept the night before, how long it has been since a day off, the level of activity, the appeal of the task at hand, the comfort of the immediate environment, the lighting levels, and so on. In addition, there is the influence of the circadian rhythm.

Circadian Mechanism

Although circadian influences will be discussed in detail in Chapter 5, they are worth mentioning at this point. The circadian drive is associated with the time of day more than with the length of continuous wakefulness. As any night worker can attest, it is very difficult to remain alert between the hours of 03:00 and 08:00, despite the fact that the prior sleep period may have ended only five or six hours earlier. Circadian factors serve to make the sleep drive particularly high in the late-night/pre-dawn hours when humans are physiologically programmed to be asleep. In addition, the circadian rhythm is responsible for the increased drowsiness that often occurs in the early afternoon, the phenomenon that has become known as the "post-lunch dip." Many people believe that this dip is a function of the fact that they have just had something to eat; however, this period of early-afternoon sleepiness occurs regardless of whether food is consumed or not. Some cultures have traditionally used this time for an afternoon siesta.

The Two Mechanisms Combined

Under normal circumstances, the homeostatic and circadian drives work together to help maintain fairly constant levels of alertness throughout the day. As shown in Figure 4.1, while circadian factors are not conducive to the greatest degree of wakefulness first thing in the morning, the homeostatic mechanism compensates for this because the preceding night of sleep has refilled the sleep reservoir and produced high alertness at this time of day. As the day wears on, the effects of the previous sleep period are gradually depleted (homeostatic mechanism), and this predisposes us to a steady decrease in alertness. However, at 18:00 or 19:00, the circadian mechanism is now quite conducive to the maintenance of wakefulness, and it opposes the homeostatic drive for sleep. If the homeostatic drive were the only factor controlling sleep/wakefulness, people would be falling asleep in the early evening. If the circadian drive were the only factor controlling sleep/wakefulness, people could not drag themselves out of bed until late morning, and they would probably fall asleep right after noon (during the "post-lunch dip"). However, primarily because of homeostatic factors, alertness is fairly high early in the morning, and primarily because of circadian factors, alertness remains sufficiently high until about 21:00 or 22:00 at night. Researchers have determined that it is actually very difficult for well-rested people to fall asleep around 21:00 at night because, while the homeostatic drive to fall asleep is high, the circadian drive is low. Of course your particular circadian factors vary; some people are morning people (larks) while some are evening people (owls). The time at which you usually wake up and the time at which you usually feel drowsy enough to return to sleep will vary with your particular circadian rhythm.

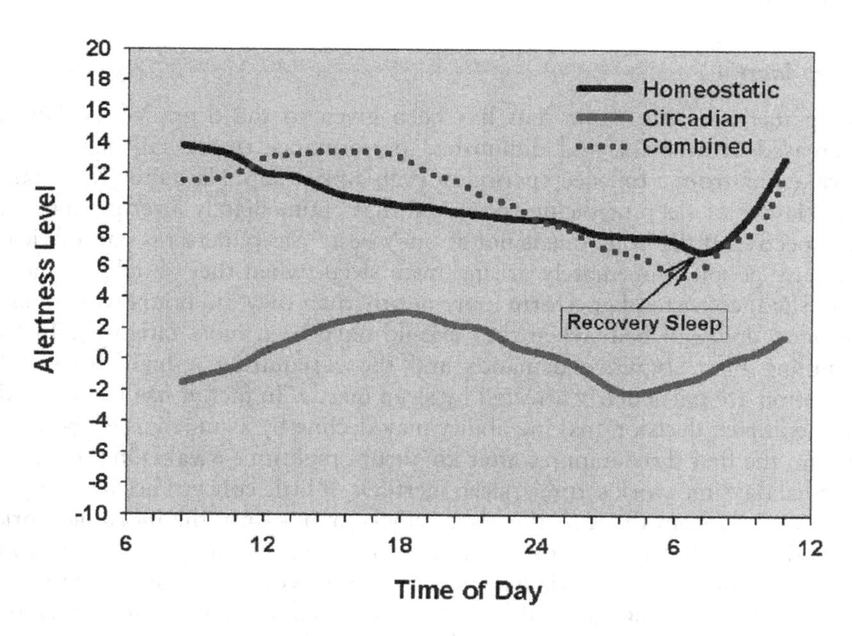

Figure 4.1 Alertness depends on homeostatic and circadian factors

Source: Reprinted with permission from Akerstedt, T. (1995), 'Work Hours, Sleepiness, and the Underlying Mechanisms,' *Journal of Sleep Research*, Vol. 4(Suppl. 2), pp. 15–22.

When both homeostatic- and circadian-driven factors begin to coincide late at night and in the pre-dawn hours of the morning, alertness levels decrease dangerously. This is when sleep becomes almost irresistible. This late-night/pre-dawn merging of the two sleep/alertness mechanisms is a serious problem for people who work non-standard schedules because their work routines require them to be awake when both of the primary underlying physiological drive mechanisms are predisposing them to fall asleep. When both the homeostatic and circadian drives for sleep are high (as is often the case during those lengthy night flights), alertness, performance, mood, and safety in the cockpit are in serious jeopardy. In fact, the combined impact of the linear decrements produced by the homeostatic drive and the cyclical degradations associated with circadian factors results in performance problems similar to those found with alcohol intoxication. One widely reported study indicated 24 hours of continuous wakefulness produced performance decrements similar to those observed with blood alcohol concentrations of 0.10 percent. Other studies have verified that performance capacity declines significantly as a function of extended wakefulness, particularly at times when both homeostatic and circadian factors are working together to promote the onset of sleep. In fact, the estimates suggest decrements of about 25–30 percent with each 24 hours of continuous wakefulness.

Sleep Inertia

Sleep inertia is the name that has been given to the degraded vigilance, increased drowsiness, and diminished performance that occur right after awakening from a full sleep period or even a brief nap. Nathaniel Kleitman, the "father of sleep medicine" observed that "immediately after getting up, irrespective of the hour, one is not at one's best." Sleep inertia is paradoxical because people immediately arising from sleep (when they should be most refreshed) consistently perform more poorly than they did hours earlier, just prior to going to bed (when they should have been most fatigued). Tasks entailing high cognitive demands and those requiring a high degree of attention are particularly affected by sleep inertia. In fact, it has been found that complex decision-making ability may decline by as much as 49 percent within the first three minutes after an abrupt nighttime awakening. In most normal daytime work settings, sleep inertia is of little concern because ample time elapses from the time the alarm clock sounds until the morning work period starts. But in aviation or in any safety-sensitive industry engaged in 24/7 operations, sleep inertia may present a serious concern if on-site napping or in-flight napping is used as an operational fatigue countermeasure, especially if skilled performance will be required immediately following the nap. Since the duration of post-nap grogginess and disorientation appears to be approximately one to 35 minutes after awakening, operational planners should conservatively allow crewmembers at least 30 minutes to fully awaken from a nap (or bunk-rest period) prior to performing duties. One last point about sleep inertia: While acute post-nap grogginess is a drawback of using naps as a fatigue countermeasure, this drawback doesn't begin to compare to the far-more-serious longer-term performance problems that are known to occur when no napping (or longer sleep) is permitted. Sleep inertia should not rule out napping as long as it is properly managed!

Other Factors

Returning to our earlier discussion about fatigue mechanisms – everyone involved in an aviation career can no doubt relate to the effects of both the homeostatic and circadian drives. Insufficient sleep logically produces increased sleepiness to signal an elevation in the drive for restorative slumber. Working during the circadian trough logically is harder than working during the normal daytime because we are genetically programmed to be asleep at this time. But is this all there is to the fatigue equation? Unfortunately, the answer to this question is no. There is a host of other considerations that determines how alert we are on the flight deck at any given time on a particular day. However, most of these other considerations relate directly to either the homeostatic or circadian factors already discussed.

One of these considerations is the amount of *cumulative fatigue* that builds up over several consecutive days of work. Cumulative fatigue is essentially a product of (1) the sleep debt that has accrued in the most recent 24-hour period; and (2) the pre-existing sleep debt that has been carried over from the preceding days of inadequate sleep. Cumulative fatigue has a noticeable impact on general alertness even though the most recent sleep episode may have provided a full eight or nine hours of sleep. Recovery from this sort of chronic fatigue may take several days to overcome. In fact, there is growing scientific evidence that five to seven consecutive days of sleep restriction produces alertness and performance decrements that persist for more than three days even when the nightly sleep period is extended to nine hours.

A second additional consideration concerns *sub-standard sleep quality* that results from environmental factors, circadian influences, sleep disorders, or psychological stress. Impaired sleep quality in aviation settings often stems from attempts to sleep at times that are not optimal from a circadian standpoint. Long-haul pilots who get stuck with the first cycle in the bunk during a flight that departs at 20:00 may get a couple of hours of sleep, but the sleep quality is certainly not equivalent to the quality of sleep obtained by the guy who gets his two hours starting at 03:00. Similarly, aviators who fly at night and try to sleep during the day would not be expected to sleep as restfully as they do when they are on normal daytime work hours with nighttime sleep hours. The timing of the sleep period can seriously affect its quality. In addition to the timing of sleep, there are other disrupters to sleep quality. Bunk sleep in the midst of noise from the galley, intercom announcements, and crew/passenger traffic is certainly less restorative than at-home sleep in a comfortable, familiar bedroom. When you are the one ultimately responsible for the flight, on-board sleep can be difficult compared to what can be obtained during a relatively stress-free weekend between duty cycles. Most captains continue to feel responsible for everything that happens on the flight deck even when there is a trusted comrade at the controls during his or her bunk cycle. The bottom line is that equivalent quantities of sleep do not necessarily have the same effect on the homeostatic drive if they are not obtained under similar circumstances.

A third additional consideration is the *"early-report-time" factor*. Having to be at the flight line at 05:00 can mean getting out of bed at 03:00 in the morning, and this is likely to be difficult from both a homeostatic and a circadian standpoint. It is very difficult for most people to get to sleep two hours earlier than usual in preparation for such an early report time due to the circadian factors already discussed. This means that sleep restriction will unavoidably result, producing an increased homeostatic sleep need. In addition, sleep quality may be affected because of concerns about starting early despite insufficient sleep. To top it all off, the requirement to roll out of bed at the ungodly hour of 03:00 in the morning will mean fighting a higher-than-normal level of post-sleep grogginess (sleep inertia)! So, early report times are a problem from all three of the mechanisms controlling alertness: homeostatic, circadian, and sleep inertia.

Fatigue-related Changes in Performance are Difficult to Predict

The degree of loss associated with extended sleep deprivation would be bad enough if it were simply straightforward, linear, and predictable. At least then it would be easy to determine when each person would likely degrade beyond "the point of no return." However, an examination of fatigue-related decrements reveals that they are modulated by temporary cyclical peaks and troughs (very much affected by the mechanisms previously discussed) that can make it impossible to predict moment-to-moment changes in operator status. Furthermore, temporary upturns in the alertness of sleep-deprived personnel (which occur primarily as a function of circadian factors) are often mistakenly interpreted as indications that some type of adaptation has occurred – a notion that will be quickly dispelled when the next trough is encountered.

Figure 4.2 depicts data that were collected from pilots flying a specially instrumented simulator during a 64-hour period of continuous wakefulness (three days and two nights without sleep). First, note that generally speaking, flight accuracy degraded from the beginning to the end of the sleep-deprivation period as would have been expected based on what we know about the homeostatic mechanism. Secondly, note that the overall performance decline was further complicated by a cyclical pattern of changes that coincide with circadian effects. There were times when performance rapidly degraded, times when performance "leveled off," and even times when

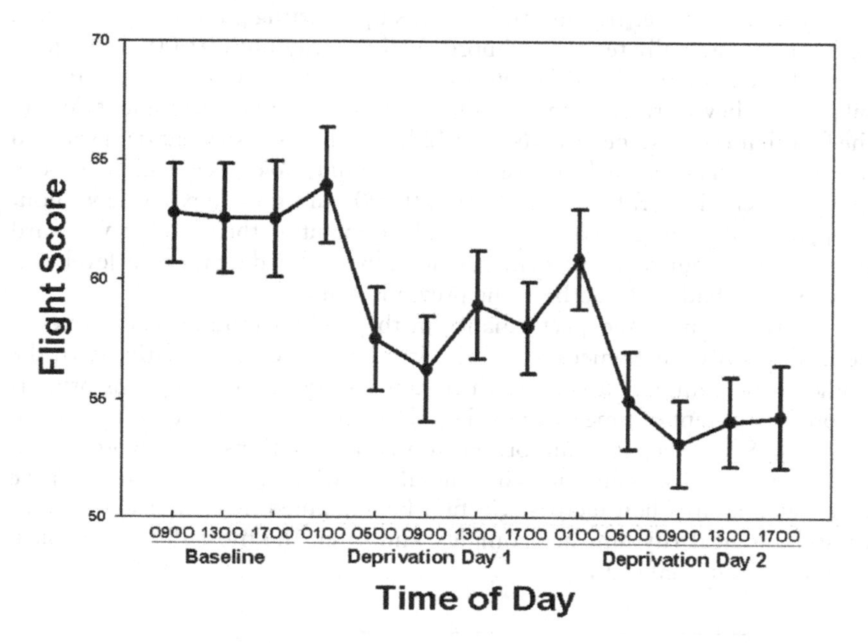

Figure 4.2 Flight performance degrades more rapidly at some times than others during sleep deprivation

Source: Reprinted with permission from Caldwell, J.A., Smyth, N.K., LeDuc, P., and Caldwell, J.L. (2000), 'Efficacy of Dexedrine for Maintaining Aviator Performance During 64 Hours of Sustained Wakefulness: A Simulator Study,' *Aviation, Space, and Environmental Medicine*, Vol. 71, pp. 7–18.

performance temporarily improved. This complex pattern of effects should not come as a surprise given what is known about the homeostatic and circadian mechanisms alluded to earlier. However, the fact that this pattern exists in flight performance, mood, and cognitive readiness raises an important point. Since fatigue-related declines are not straightforward and not linear, they can be very difficult to predict. Notice in the graph that from 09:00 on the baseline day (after a good night's sleep) until 01:00 on the first deprivation day, performance remained fairly consistent despite 18 hours of continuous wakefulness. Of course, the pilots being tested were very well rested on this first day, and as they were becoming more familiar with the flight profile, their increased familiarity with the flight tasks tended to offset some of the fatigue-related decrements. Because of this, by 01:00, many of them were thinking "hey, I can go on like this forever without having to worry about fatigue!" But notice that between 01:00 and 05:00, with the addition of only four more hours of sleep loss, the performance decrement made it seem as if the time had been much longer. Then from 05:00 to 09:00 there is yet another decline – just at that time when some of the aviators thought they would get better because the sun was up. Starting later in the morning, performance actually tended to improve all the way until 01:00 on the next day! This circadian-related benefit often created the impression among the pilots that they were adapting to fatigue and learning to overcome it despite the fact that they had been awake for 42 hours. But, subsequent data showed just how dangerous such an assumption would have been had they been flying an actual aircraft. From 01:00 to 09:00 during the next cycle without sleep, not only were the volunteers unable to maintain their tendency toward improvement, but their performance actually declined again to a level that was twice as bad as it had been the previous night.

Clearly, alertness and performance in the cockpit (and everywhere else) depend heavily on homeostatic and circadian processes. Of the two, the homeostatic component is more easily understood since it depends primarily on the amount of time since the last sleep episode and the quality of sleep obtained. Since sleep is so important, subsequent sections will devote a good deal of time to discussing sleep in general and what can be done to improve it. However, circadian factors will first be explained in greater detail since these biological rhythms are complex and often the most difficult to deal with in operational settings.

Top Ten Points About the Processes Underlying Sleepiness and Alertness

- The primary determinant of the level of fatigue is the time awake since the last sleep period.
- The next determinant of fatigue is the time of day according to the body's internal clock.
- The third major factor is the amount of post-sleep grogginess upon awakening.
- These three factors are called the homeostatic, circadian, and sleep-inertia processes.
- On a consistent day schedule, the homeostatic/circadian factors help maintain consistent alertness levels during daytime hours.
- On night duty, circadian-induced fatigue is most pronounced from 03:00–05:00.
- On a night schedule, this circadian effect exacerbates the effects of being awake for a long time.
- Several days of shortened sleep cause an accumulation of fatigue from which it is difficult to recover.
- Sleep disturbances that impair the restorative value of sleep further impair alertness levels.
- Trying to predict the exact effects of fatigue is difficult due to wide variation in moment-to-moment decrements in mental abilities.

5 Circadian Rhythms

Each of us has an internal mechanism called a biological clock that affects the biological and psychological processes that naturally vary over the 24-hour day. These processes are termed "circadian rhythms" from the Latin "circa" meaning *about* and "dias" meaning *day*. So circadian rhythm means a pattern that varies on a cycle of approximately 24 hours. Examples of some of the processes are body temperature, hormone secretions, and alertness, but it is important to recognize that there are many more processes within us that vary coincident with our 24-hour day. There are a number of cues, called "zeitgebers" (German for "time cues") within the environment, which help keep these processes in sync, and when everything is in sync, we feel pretty good. Light is the primary zeitgeber which keeps the rhythms consistent (locked to a certain time of day) and in synchrony with one another, but there are other zeitgebers such as social factors (meals, work activity, and so on) which contribute to the stability of these internal processes as well.

The work of Dr Nathaniel Kleitman, the father of sleep science, was instrumental in helping us understand internal body rhythms. In 1938, Dr Kleitman and another researcher spent about a month in Mammoth Cave, Kentucky, isolated from all external timing cues. The only light source was from artificial light, which was controlled by a fixed 28-hour schedule. The researchers were interested in setting their days to 19 hours of activity and nine hours of sleep rather than the more typical 16 hours of activity and eight hours of sleep. They tried this for 30 days, during which they recorded body temperature and activity level throughout the experiment. The result was that only one of the two guys came close to adjusting to the 28-hour day, while the other did not adjust at all. Both became sleepy at their "normal" bedtime, but only one of the researchers' body-temperature rhythm (one of the best measures of a person's body clock) adjusted to the new sleep/wake cycle.

Other investigators have attempted to change the body cycle to something different from 24 hours, and the ones that were most successful were the ones who chose new rhythms that were close to the standard 24 hours (22 to 26 hours), whereas those who selected rhythms that were much different achieved poorer results. The bottom line – people have trouble adjusting to new cyclical patterns of wakefulness and sleep that do not coincide with the

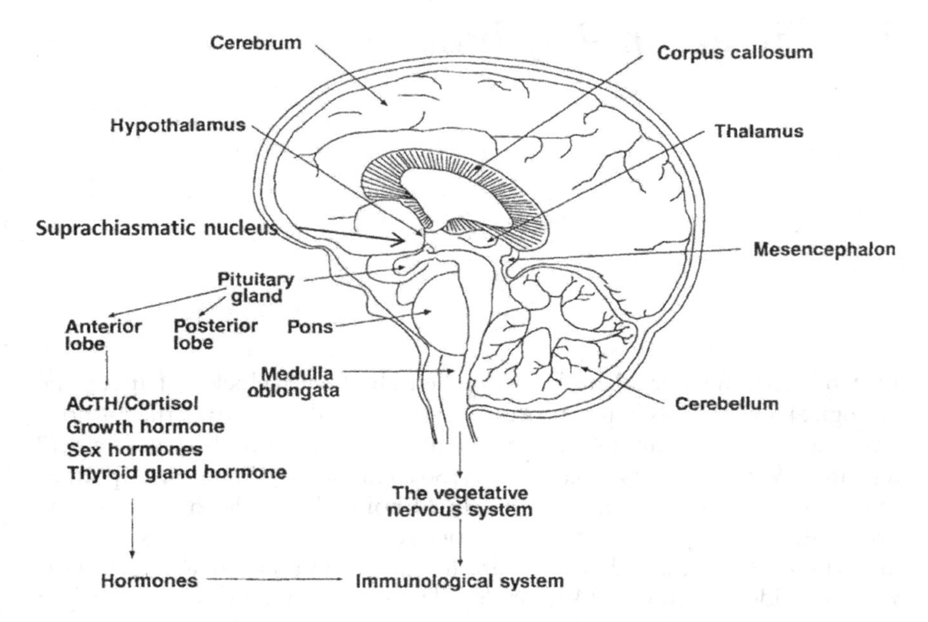

Figure 5.1 The location of the SCN that synchronizes the body rhythms

Source: Reprinted with permission from Advisory Group for Aerospace Research and Development, North Atlantic Treaty Organization (AGARD/NATO), (1994), 'Psycho-physiological Assessment Methods,' AGARD Advisory Report, AR-324, p. 22.

normal 24-hour Earth day even when they establish almost perfect conditions to facilitate the change.

Why is this adjustment so difficult? The main reason is that the human brain is "hard-wired" for approximately a 24-hour day. It is programmed to be asleep during the dark hours when the sun is down, and awake and active during the daylight hours when the sun is up. Remember our earlier discussion about how historically adaptive it has been for human beings to live on this sort of schedule? For thousands of years, people remained in the same time zone most (if not all) of their lives, and they rested at night and worked during the day simply as a matter of practicality. It is hard to forage and avoid predators when you cannot see where you are going. So when it is dark, the best course of action is to stay safely inside of your cave!

With the Earth's 24-hour rotation, the cycle of light and dark (at least in most places on the planet) keeps us on this 24-hour day. Physiologically, this happens because light, perceived through the eyes via the retina, is transmitted to the brain's biological clock – the suprachiasmatic nucleus (SCN). The SCN, which is located just above the optic chiasm and in front of the hypothalamus, receives the light stimulation through the optic nerves (see Figure 5.1). Through a series of other nerves, the SCN is linked to a variety

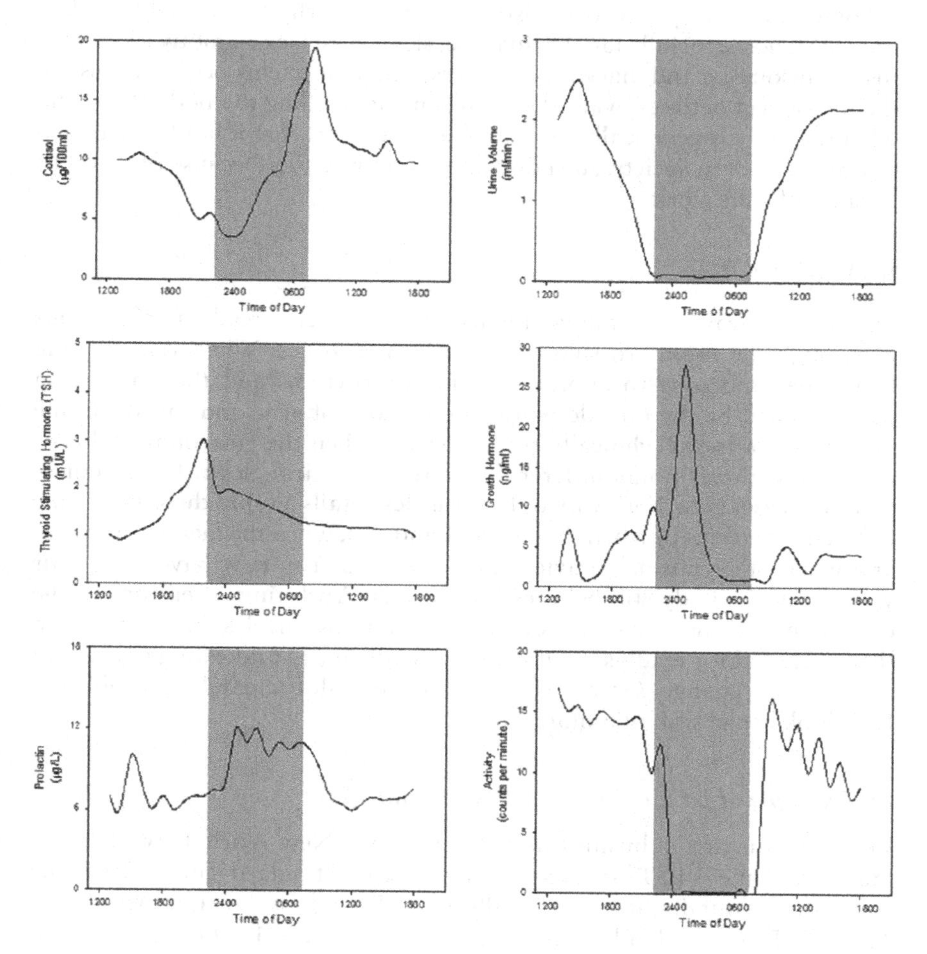

Figure 5.2 A small subset of the body's internal rhythms

of regions in the brain that control behavior. In concert with the light stimulation, the SCN makes sure all the body's rhythms are synchronized to a 24-hour cycle.

When everything is constant and normal, the body's rhythms work together in harmony like an orchestra (and there are plenty of rhythms inside of the body, as shown in Figure 5.2). Although at any given minute of the day, some rhythms are up and others are down, the moment-to-moment fluctuations occur harmoniously, in tune with one another, to keep the body running smoothly. But when something disrupts the body's time schedule, problems with this smooth orchestration of rhythms rapidly begin to occur.

Some of the symptoms of a disrupted circadian rhythm, called *circadian desynchronosis*, include fatigue, malaise, sleepiness, lack of motivation, confusion, insomnia, and digestive disorders. These problems occur because of the disconnect between what the environment is telling the body to do and what the body is genetically programmed to do (or what it has been trained to do). In modern society, such disconnects occur either because of travel or because of shift work.

Jet Lag

One of the ways that pilots frequently encounter circadian disruptions is through the rapid crossing of multiple time zones. Whenever traveling from one time zone to another, the body's rhythms and the time-giving cues supplied by the outside world become out of sync and the syndrome of "jet lag" arises. Technically, jet lag occurs when the environmental cues are at least threes hours different from the biological clock. For example, traveling from New York City to Los Angeles entails a rapid three-hour time difference in terms of when the sun rises and sets, when the meals are served, and when other normal activities are scheduled. You may have reset your watch, but the body still believes it is on New York time, even though the environmental cues are saying everything is on Los Angeles time. Obviously, there is a need for readjustment here, but surely it can't take long to get over a three-hour change. Or will such a small time shift create any problems? Let's look at the situation more closely.

An Example of Jet Lag in an East/West Transition

Imagine you are a businessman flying from New York City (Eastern Standard Time – EST) to Los Angeles (Pacific Standard Time – PST) for a meeting. Your flight leaves at 09:00 EST, to arrive in Los Angeles at 12:00 PST. As you land, your body's clock says it is 15:00, but clocks in Los Angeles say it is noon. No problem. You'll just eat lunch, find your hotel, and meet some colleagues. You arrive at your hotel and settle in. Looking at your watch, it is 17:00. Your body says it is 20:00. You are ready for dinner, but your dinner partner who lives in Los Angeles says it's two hours too early. Finally, at 19:30 PST you meet for dinner. Local time says it is 19:30, but your stomach says it is 22:30 so you are starved! You eat a big meal, engage in a little small talk, and then call it a night. By the time you return to your hotel, it's only 21:00 local time, so you decide you'll watch a little TV, read for a while, or maybe even go for a walk. However, your body has other plans. The alertness monitor in your brain (remember the circadian component?) started down a long time ago; it is midnight by your brain's clock and you are really tired. So you hit the sack. Sleep is no problem; you go right off. However, you wake up at your "normal" time which is 07:00 EST. You feel fairly alert and ready to get up, but the clock in your hotel room says it is only 04:00! OK, so you just have to go back to sleep. But this is easier said than done since your circadian cycle is now on its way up and you have managed to get at least a few hours of restful sleep. Finally, at 05:30 PST you give up on sleeping any longer, get out of bed, and just try to amuse yourself until it's time for the 09:00 meeting.

So what problems occurred when moving the environmental clock back three hours? First, mealtime was very late in terms of the biological clock, and bedtime was pushed back several hours as well. Usually this is not a significant problem. The body's "free running" natural rhythm (the rhythm it would assume in the absence of any time cues) tends to run slightly longer than 24 hours, so extending the day actually agrees somewhat with the body clock's natural rhythm. Exposure to the new environmental zeitgebers (later sunset, later dinner, and later activity) also helps to more rapidly establish synchrony between the environmental clock and the body's clock. The main problem is waking up too early in the morning, which leads to sleep restriction. Remember that the late bedtime was unfortunately not accompanied by a later wakeup time, and this shortened the sleep period by at least a couple of hours. Research tells us that even this amount of acute sleep loss will make us feel pretty tired later in the day. However, if the trip is only a couple of days long, even this will not be much of a problem. So overall, this westward time shift really was not too tough.

An Example of Jet Lag in a West/East Transition

But what about when your associate who works on the west coast flies from Los Angeles to a meeting in New York City? Let's go through this scenario. The flight leaves at 08:30 PST from Los Angeles, landing in New York City at 16:30 EST. His body clock at landing is 13:30. You plan to meet him for dinner at 19:00. When 19:00 finally comes, you are ready to eat, but he is barely hungry because his stomach thinks it is only 16:00. But he eats a light dinner with you to be polite before returning to the hotel. He watches TV for a while waiting for bedtime. He doesn't feel very sleepy, but the clock on the nightstand says it is 23:00. Unfortunately, his brain thinks it is 20:00. He knows he needs to get some sleep, but no matter how hard he tries, he can't seem to force it. But finally, at about 01:00 he falls asleep. He's thinking this is a lot later than he usually gets to sleep, but it actually coincides with his body's usual time of about 22:00. He has set the alarm for 07:00 in order to make your 09:00 meeting. This seems like plenty of time to sleep, but when the alarm sounds, he can barely open his eyes. He has been asleep for six hours, which is not too bad under the circumstances, but he is trying to drag himself out of bed when his brain thinks it is 04:00! As discussed earlier, this is a really low point in the circadian rhythm so it is pretty tough to hop out of bed at this time of day. But he has to get up because the meetings are important.

So what problems occurred for your friend when the environmental clock was advanced (rather than delayed) by three hours? First of all, it was very difficult to go to sleep earlier than the body's usual schedule because this put the start of the sleep period at a time prior to the opening of what has been called the "sleep gate" (a sleep gate is the window of time in which you can most readily fall asleep). The body's chemicals and internal activation levels are not set to allow sleep to come early (before the "gate" opens), so it is really hard to get to sleep before your normal bedtime unless you are sleep deprived. Secondly, it was very difficult to get out of bed at that earlier-than-normal wakeup time because it is hard to overcome that sleepy, muddled feeling, called "sleep inertia", that is especially noticeable when trying to wake up during the circadian trough.

Of course, as was the case when you traveled to Los Angeles, your friend will benefit from exposure to the environmental zeitgebers such as the changes in the light/dark cycle, meal times, activity rhythms, and so on. The exposure to these cues will aid adjustment to the new environment. However, your friend will experience more trouble readjusting than you did because he is making an eastward transition (compared to your westward transition). This means that his first day was shorter than usual whereas your day was longer than usual, and as noted earlier, the body more easily adjusts to *delays* than to *advances* in bedtime. So, it will take a few days longer for your associate to accomplish a complete resynchronization, and meanwhile, his body will respond to its timing disruption with indigestion, headaches, sleepiness, and insomnia until all (or most) of the internal rhythms again agree with the external environmental time-of-day markers.

How Long Does it Take to Get Over a Case of Jet Lag?

Research indicates that someone can adapt to a five- to eight-hour time change in as little as two days, but it more often takes as long as two weeks. The rule of thumb is that it takes about one day of adjustment for every hour of time zone change; however, this will depend on the direction of the time change (westward versus eastward) and the degree of exposure to environmental cues. As discussed earlier, the body is able to lengthen its internal day relatively easily, and therefore, it can adapt to a westward time change (when the day is lengthened) faster than it can adapt to an eastward time change (when the day is shortened). Note from the diagram (Figure 5.3) how eastward travel requires early sleep initiation and early awakening, both of which are difficult for the body to accommodate. However, westward travel presents a situation that is more like what many of us already do on our off-duty time – staying up later at night and sleeping later the next morning.

The scenarios discussed earlier demonstrate only mild cases of jet lag since there is only a three-hour disagreement between the environmental zeitgebers and the biological clock. However, they help to illustrate the

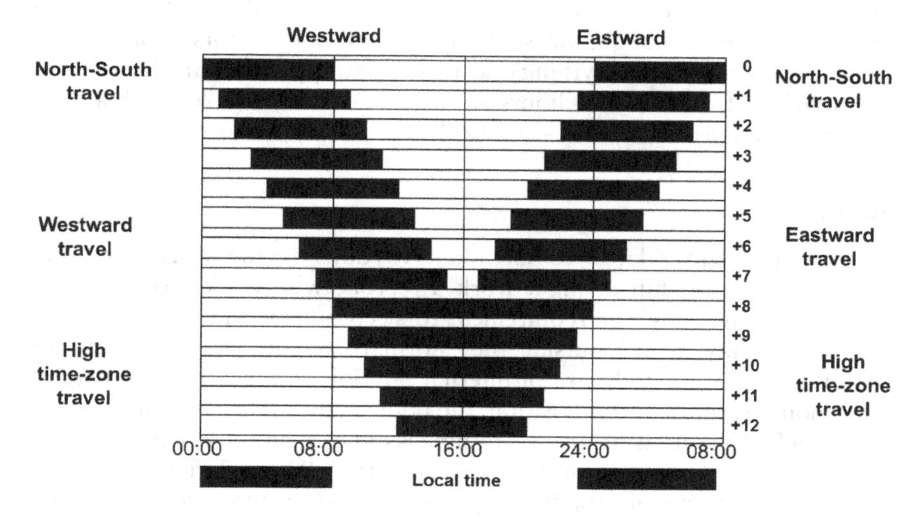

Figure 5.3 Time changes following eastward and westward flights compared to the "normal" time

Source: Reprinted with permission from S. Bourgeois-Bougrine and S. Folkard.

nature of the more serious problems that occurs when flying transoceanic routes that lead to six to nine-hour disconnects between the biological clock and the environmental clock.

Jet Lag is a Continuous Problem for Long-haul Pilots

For travelers who are vacationing or working at their new destination for a week or more, jet lag is always problematic, but only for a few days. They will soon be able to readjust to their new sleep/wake schedules since they routinely will be exposed to a consistent new set of environmental cues. Pilots and flight crews who frequently make this type of transition are not as lucky because they rarely stay within the new time zone long enough for readaptation to occur. Like the travelers, they suffer from alertness decrements as a result of disruptions in their circadian rhythms (circadian desynchronosis) as well as from sleep loss associated with an inability to initiate and maintain sleep during designated layover periods. However, unlike the travelers, they are off flying to a new destination (complete with a new set of environmental time cues and work schedules) by the time their body clocks begin to shift to the schedule they just left. No wonder routinely flying international routes can produce a state of chronic jet lag that compromises the ability to stay awake and alert in the cockpit! And in-flight environmental factors do not help the situation since the pilot's physical activity is restricted, there is little control over environmental light cues, and the highly automated cockpit is not particularly stimulating. Commercial pilots have it bad enough, but some of our military aviators really have it tough since, on top of everything else, they are strapped into an ejection seat for anywhere from 12 to 30 hours at a time. There is no galley with a fresh pot of coffee brewing on an F-22 or a B-2!

Shift Lag

Another problem related to circadian rhythms is the disruption that occurs because of shift or night work. Everyone who works non-standard schedules knows how the requirement to sleep outside the "normal" sleep period and stay awake outside the "normal" activity period affects the smoothly operating orchestra of internal physiological rhythms. Instead of a harmonious musical composition, the body sounds more like the tune-up period before the show begins; everything is out of sync and discordant. This uncomfortable circumstance is called "shift lag." While shift lag produces the same symptoms of fatigue, sleepiness, and so on that occur in jet lag, the problems usually last longer since the environmental cues are often in opposition to optimal circadian rhythms.

An Example of Shift Lag Associated with Night Work

As an example, consider night work. While the body is being told to sleep during the day and stay awake during the night, the environmental cues are saying just the opposite. Light is coming in to alert our brains to begin a new day just like it has been helping our ancestors to wake up for thousands of years. Unfortunately, our brains are getting this powerful cue when we are on our way home to get some sleep. Not only is the sun not cooperating, but friends and family are getting ready to go to work when we are going to bed. Then, while we are leaving the house to start the next shift, it is already dark and everyone else is going to sleep. From an evolutionary standpoint, environmental lighting is again working against us, and on top of that, the social cues are saying it is time to get done with this day. Just from what the surrounding environment is telling us, it is no wonder it's so difficult to be a shift worker!

Discordance Between Schedules and Environmental Cues is a Major Problem

Not only are the social cues a real hindrance to night-shift adaptation, but there are physiological factors as well that make non-standard sleep/wake schedules a real pain. Sleep is very difficult to obtain during the daytime hours for several reasons. First of all, sunlight exposure excites the neurons of the body clock to signal the beginning of the new day (that is, to wake up). Body temperature rises, and hormonal levels are shifting to mobilize energy for the body and the brain. While sleep deprivation from staying awake the preceding night (and often the preceding day) promotes rapid sleep onset at the end of the night shift, circadian factors disrupt sleep maintenance as soon as some of the sleep pressure has subsided. After about three to four hours of sleep, most night workers awaken and find it very difficult to return to sleep and remain asleep for very long. Most give up after a while and decide to engage in normal daytime activities until it is time once again to report for work at night. The fact that these workers expose themselves to light and activity when they would like to be asleep makes the long-term situation worse because these cues keep the body's clock on a daytime waking schedule even longer, making it hard to sleep during the day tomorrow. The cycle continues until either fatigue sets in so heavily that the worker eventually collapses into sleep (hopefully while at home), or the night-work cycle ends, permitting a return to typical daytime work and nighttime sleep.

The Timing of Sleep Periods is Critical for Sleep Quantity and Quality

Much of the ability to sleep during the day depends on the time at which sleep is initiated. Generally, when a person is able to go to bed and try to fall asleep in the early-morning hours, for example before 06:00, then it is relatively easy to fall asleep. However, waiting until the mid to late morning makes sleep more difficult to initiate and maintain. Figure 5.4 depicts the findings of various sleep experts regarding the ease of falling asleep at

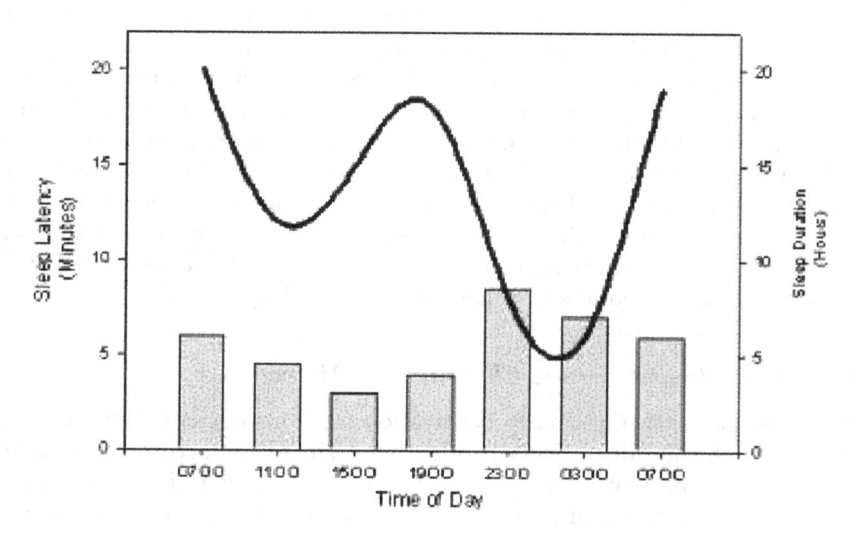

Figure 5.4 The mean time for individuals to fall asleep by time of day (sleep latency is depicted by the solid line; mean sleep duration is depicted by the bars)

different times of day over the 24-hour period (solid line). Remember that these are average times and individual differences should be taken into account. The figure also shows the average length of sleep obtained as a function of the time of day when the sleep was initiated. Generally speaking, it is easy to see that day sleep is not only harder to initiate, but it tends to be shorter than night sleep. This makes it very difficult for night workers to obtain the recommended eight hours of sleep. Note that if someone tries to sleep at 11:00 in the morning, it may take only 10 to 15 minutes to go to sleep (because of that high homeostatic drive we discussed earlier), but the sleep only lasts on average less than five hours, primarily because the circadian factors (also discussed earlier) are not conducive to sleeping at 17:00 or 18:00. However, when starting to sleep at 23:00 at night, it takes on average less than ten minutes to fall asleep, and the sleep duration is about eight hours. Both the homeostatic and the circadian pressure to sleep are strong at 23:00 at night, but the circadian pressure for sleep is very weak at 11:00 in the morning.

The Timing of Work Periods Affects Alertness on the Job

It is easy to see that getting enough sleep during the day is difficult, but staying alert during the night is also a problem. Again, the night worker's zeitgebers are saying rest and sleep, while the work setting is saying it is time to wake up and do your job. The darkness following sunset is a powerful cue for sleep. Other people are slowing down and preparing to call it quits for the day, the hormones in the body are preparing the brain for sleep, and body temperature is falling in preparation for the upcoming state of inactivity. All of these factors make it very difficult for people to stay alert at night for more than a few hours before the sleepiness becomes overwhelming between 02:00 and 06:00. For most of us, it is a real battle between our willpower and our physiology to try to remain alert at this time of night, and the result is reflected in the number of unintended sleep episodes that occur. As Figure 5.5 indicates, involuntary sleep lapses are much more frequent in the early-morning hours than at other times of the day. In addition, a circadian dip in the early afternoon is also associated with an increase in the frequency of these unintended sleep episodes.

Night Work Presents Serious Challenges for Aircrew

As mentioned earlier, the sleep deprivation from insufficient daytime sleep and the nighttime drowsiness produced by circadian factors combine to create significant problems for aircrew who are working nights. Accidents are more likely, errors in performance are more frequent, mood is poorer, and motivation is lower at night than during the day. Night work not only poses safety risks, but it seems to impair immune-system functioning as well, making those on the night shift more prone to illness than their daytime

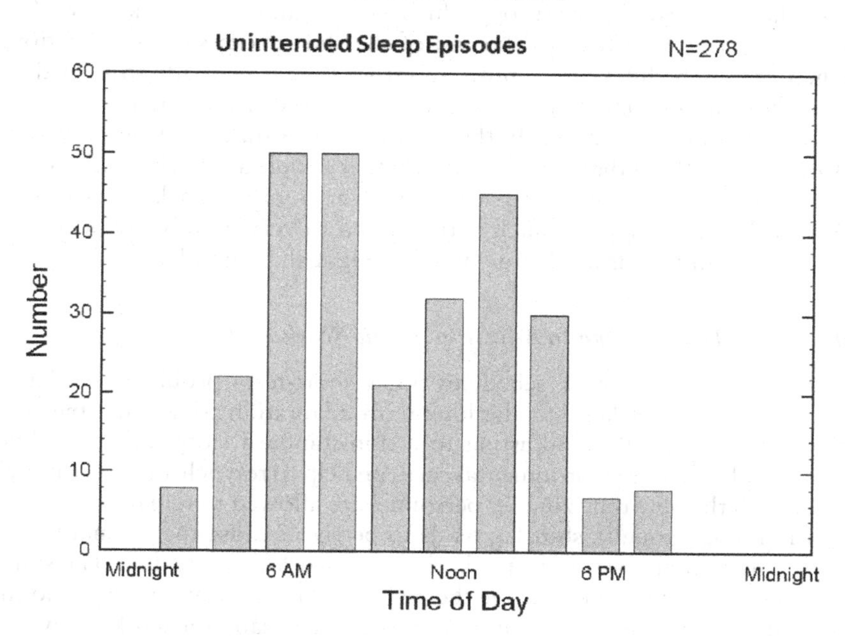

Figure 5.5 The number of unintentional sleep episodes by time of day

Source: Reprinted with permission from Mitler, M.M., Carskadon, M.A., Czeisler, C.A., Dement, W.C., Dinges, D.F., and Graeber, R.C. (1988), 'Catastrophies, Sleep, and Public Policy: Consensus Report,' *Sleep*, Vol. 11, pp. 100–109.

counterparts. Outside of work, the fatigue from the night shift leads to high levels of sleepiness during the drive home, and as a result, vehicular accident rates are particularly high for workers driving home after a night of work. All in all, shift work creates a variety of difficulties in terms of performance, safety, and general well-being.

Early Report Times are Problematic as Well

By the way, although the effects of early shifts are commonly downplayed by schedulers, crews who must show up at Operations early in the day (before 06:00–07:00) can experience alertness problems similar to the ones encountered by those who are stuck with the "red-eye" flights. Although the personnel working early shifts have the advantage of being on the job during daylight hours (a significant plus), they have the disadvantage of being sleep deprived from inadequate pre-mission sleep due to circadian factors. Many people experience a great deal of difficulty falling asleep early in the evening (before their sleep gate opens), which leads to insufficient sleep duration. Remember our business traveler from earlier? These folks have problems waking up at 04:00 because this is the low point in their circadian cycle. So, they are groggy during the beginning of work (from sleep inertia) and drowsy toward the end of the shift (from sleep deprivation). When early schedules are a must, an effort should be made to staff them with those individuals who, through genetic predisposition, are naturally more alert in the morning. By the way, there is a quick test developed by Horne and Östberg that can identify whether people are Larks (early to bed and early to rise) or Owls (late sleepers who are more active late in the day). Obviously, the Larks are much better suited to very early work schedules than Owls, and Owls are better suited to night shift than Larks.

How Long Does it Take to Adjust to a New Work Shift?

Adaptation to new work schedules poses short-term problems similar to those caused by jet lag, but the long-term adaptation to a new time zone is generally easier than adjusting to a non-standard sleep/wake schedule. Unlike jet lag where environmental cues can help to resynchronize the body's clock with the new time zone (if personnel are allowed to remain in the new location long enough), shift lag tends to persist because the normal timing cues remain in opposition to the body's new work/rest schedule. The sun is out and the world is active while the night worker is trying to sleep, and the opposite is true while the night worker is getting ready for work. However, if every effort is made to facilitate adjustment to a new work schedule, the amount of time to a reasonable adaptation is about one full day for every hour of schedule change.

Are Slow Schedule Rotations Better Than Fast Schedule Rotations?

Careful planning of schedules will assist in adjustment to night duty, but complete adjustment is difficult (and some say it is impossible). This partially explains why there are different schools of thought on the scheduling of shift work. One suggests that shift rotations should be kept to short two to three day durations (rapid rotation), while the other suggests that any new work shift should be maintained for at least five to seven days (slow rotation). Support for the rapid rotation strategy comes from the fact that full adaptation to the shift change is unlikely anyway, and the longer someone remains on the night shift, the higher his cumulative sleep debt will become. Proponents of this strategy feel that keeping the night rotation short will prevent fatigue from building to dangerous levels before a normal day shift is resumed. Support for the slow rotation strategy is based on the fact that people can at least partially adjust to the new schedule if they use appropriate techniques and if they remain on the new schedule long enough. Although short-term adaptation problems may occur, in the long run, alertness will improve as the circadian rhythm adjusts, and the ability to obtain adequate daytime sleep improves.

The Bottom Line

The bottom line is that the body will feel like it is revolting whenever the circadian rhythm is disrupted. While there are strategies to help adjust to new time zones and new work schedules, the circadian system is fragile and it will complain whenever it is striving to synchronize to a new schedule of external environmental time cues. Strategies for maximizing adjustment are discussed later in this book, and adhering to these suggestions will minimize the amount and duration of discomfort from circadian disruptions. However, even in the best of circumstances, the adjustment to any new schedule will take time and, until readaptation occurs, mood and performance will be compromised. Since many of the problems will stem directly from disrupted and insufficient sleep, it is very important to focus on minimizing anything that will further exacerbate the sleep loss. As the next chapter will show, sleep is a surprisingly complicated process.

Top Ten Points About Circadian Rhythms

- Everyone has internal, physiological 24-hour cycles called circadian rhythms.
- The Earth's light/dark cycle maintains the day-to-day consistency of these rhythms.
- Time zone or work/sleep-schedule changes desynchronize rhythms, causing fatigue and other problems.
- Jet lag occurs when time zone crossings lead to conflicts between the body's clock and the new environmental time cues.
- It takes a minimum of one full day (at best) to adjust to every one-hour time zone shift.
- It is easier to readjust to time changes after westward travel than eastward travel.
- Shift lag occurs when altered work/sleep schedules create a disparity between the body's clock and the environmental time signals.
- Shift lag is hard to overcome because shift workers are constantly fighting an opposite environmental light/dark cycle.
- Problems performing while jet lagged or shift lagged are made worse by poor off-duty sleep.
- Constantly changing schedules or time zones will increase the levels of cumulative fatigue.

6 Sleep Facts

By now it is obvious that the best anti-fatigue measure is to ensure sufficient quality sleep on a daily basis. This will take care of your homeostatic sleep drive, especially if you can supplement plenty of consolidated off-duty sleep with a nap. It will not directly affect your circadian drive, but it will reduce the impact of the circadian trough by attenuating the additive effects of the homeostatic drive. Besides sleep, nothing else can do more to enhance on-the-job alertness, ensure optimal judgment and concentration, and promote a healthy, positive outlook. Your objective should be to obtain the consistent average of seven to eight hours of sleep per day that experts recommend. Even if shift lag or jet lag impairs your ability to acquire eight consolidated hours, you should supplement with naps in an effort to get eight *total* hours. By the way, you might one day decide that you are one of those lucky people who can get by on less than this recommended amount of sleep, or you might decide you are unlucky enough to need more time in the sack every day. However, until you become absolutely certain that you can be your best on less than eight hours a day, or until you become convinced that you really could use more than eight hours, stick with the recommended amount. You will know that you are sleeping enough when you don't answer "yes" to most of the following questions:

- Do you fall asleep in under five minutes after going to bed?
- Do you always feel like you could take a nap?
- Do you become drowsy after eating a big meal?
- Do you fall asleep when watching TV, sitting in meetings, or when otherwise sitting still?
- Do boring activities make you sleepy?
- Do you sleep an hour or two longer than usual on days off?
- Do you find that you can hardly make it through work without caffeine?
- Do your eyelids feel "heavy" or droopy while you're at work?
- Do you find that your head nods periodically while flying or performing those boring administrative tasks?

If you answered "yes" to one or more of these questions, you probably are not getting enough sleep to keep you at your best every day. This could mean that you have a sleep disorder or some type of transient sleep problem, that you just need to reprioritize the importance of sleep versus other activities in your life, or that you are suffering from insomnia due to poor sleep habits. We will discuss each of these issues in turn, but first it is worthwhile to spend some time on a general discussion on the nature of sleep itself.

The Process Called Sleep

What happens when we go to sleep? Does the brain just lie idle, waiting for a noise to wake it up so that it can begin working again? The answer to this question is no. Far from being idle, the brain is very predictably active during sleep, and as it turns out, most of us have very similar sleep patterns every night.

Understanding the Process of Sleep Through Physiological Recordings

In order to see this activity called sleep and to quantify the changes that occur in the transition from wakefulness to sleep, researchers have for years been putting recording sensors on the scalp, near the eyes, and under the chin, and connecting these sensors to very sensitive amplification equipment so that they can study sleep. These sensors and amplifiers monitor brain activity (electroencephalograph – EEG), eye activity (electroocculograph – EOG), and muscle activity (electromyograph – EMG), and they write the results (a series of undulating lines) out on paper or in a computer where they can be further examined and classified. Usually, brain activity is recorded from several sites on the scalp. Many sleep researchers use a couple of central recording sites – one on the top-right side of the head and one on the top-left side of the head; and a couple of occipital recording sites – one on the right of the back of the head and one on the left. Likewise, there are usually two eye monitoring sites – one near the left eye and one near the right eye. The muscle activity is recorded from underneath the chin. In 2012, the American Association of Sleep Medicine revised the guide to conducting sleep recording and scoring, adding two EEG sites (F3 and F4), combining slow-wave sleep into one stage rather than two, and including information to address the digital technology used in modern sleep laboratories. The diagrams in Figure 6.1 show the placement sites for these sensors.

Classifying the Different Types of Sleep

For many years, all of the squiggly EEG, EOG, and EMG lines that make up a sleep recording (or polysomnogram) seemed to be just a chaotic mess. But thanks to Dr Nathaniel Kleitman (the guy who did the cave studies), Dr William Dement (one of Dr Kleitman's students), and some others, it soon became evident that there was order in what at first appeared to be chaos. Then in 1968, Dr Allan Rechtschaffen and Dr Anthony Kales published a guide that provided a standardized way to classify sleep into specific stages.

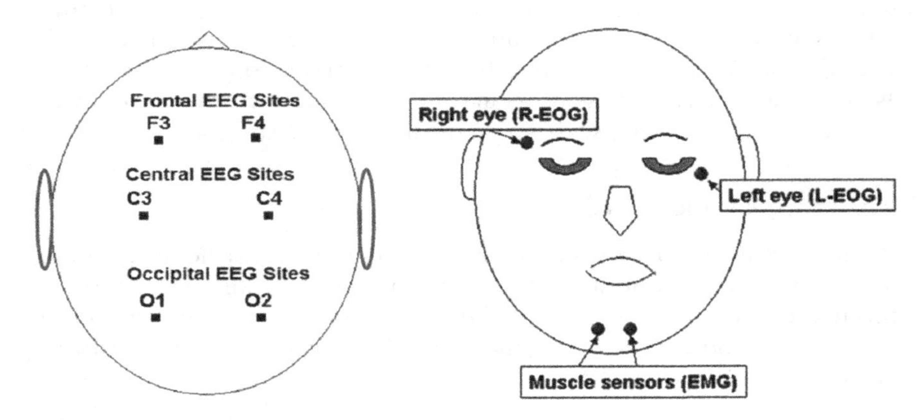

Figure 6.1 Sensor placement sites for EEG, EOG, and EMG recordings

This suddenly made it possible to precisely compare sleep recordings made by professionals around the world, to pool the information collected from many different laboratories, and ultimately, to advance the field of study to the point that sleep and sleep disturbances could be better understood. By examining a person's sleep record, it is possible to determine whether he is awake or asleep, to know exactly when he fell asleep and woke up, and to differentiate between different types or stages of sleep.

Non-REM and REM Sleep

Sleep is separated into two distinct states – non-rapid eye movement sleep (NREM) and rapid eye movement sleep (REM). NREM sleep traditionally has been further divided into four stages that progress from the lightest – Stage 1 sleep – to the deepest – Stage 4 sleep. Each of these stages is discussed below. As mentioned above, clinical labs are beginning to use the new staging rules which divide sleep into 3 NREM stages – Stage N1 (same as Stage 1), N2 (same as Stage 2), and N3 (combines Stages 3 and 4 into one slow-wave sleep stage). However, we will discuss sleep in terms of the traditional 4 NREM stages.

Awake

When we are awake and active, our EEG activity is fast and desynchronized. The name that has been given to these fast EEG waveforms is *beta activity* (these waveforms fluctuate at a speed of over 12 cycles per second). When we close our eyes and relax, the EEG slows slightly to a pattern of activity ranging between eight and 12 cycles per second called *alpha activity*. Alpha is more uniform and synchronous or rhythmic than beta. Usually, when a

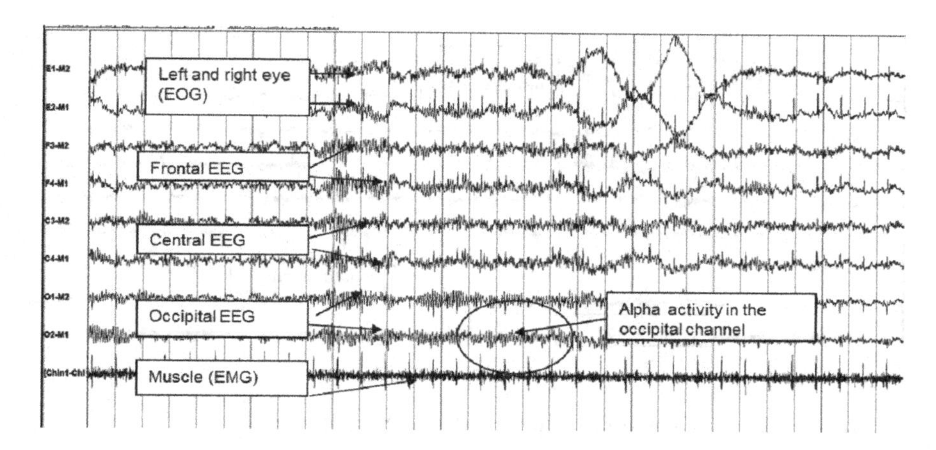

Figure 6.2 EEG, EOG, and EMG activity during wake

person relaxes, the eyes become relatively still (since they are closed) and overall muscle tension is reduced. However, a relaxed person is not necessarily asleep, and the patterns observed in brain, eye, and muscle activity reflect this fact. Figure 6.2 shows the EEG, EOG, and EMG patterns which normally occur when a person is relaxed, but awake, with the eyes closed. This record is more than 50 percent alpha activity with eyes closed; EMG is high; EOG is relatively inactive. The figure represents a 30-second epoch of data.

Stage 1 Sleep

When we lie down and begin to fall asleep, the brain activity slows even more than it did under conditions of relaxation. This slower EEG pattern is called *theta activity*, and it is about four to eight cycles per second. When a polysomnographer (someone who interprets sleep records) sees a combination of alpha and theta activity, with the majority of the activity in the theta range, the activity is labeled Stage 1 sleep. If we look at the eye recordings, the EOG pattern, there are slow movements as the eyes roll from side to side in a pendulum fashion. This is another indication of Stage 1 sleep. However, note that muscle activity remains relatively the same as it was under conditions of relaxation (indicating the presence of muscle tone) even as the other features suggest sleep onset. Figure 6.3 shows the EEG, EOG, and EMG patterns typical for Stage 1 sleep. This epoch of sleep represents 30 seconds of data.

Stage 1 sleep is the transition between wake and sleep. During this stage, there may still be some awareness of activity in the surrounding environment. Conversations might still be heard even though the eyes are closed and the brain is transitioning into sleep. Someone who is awakened from

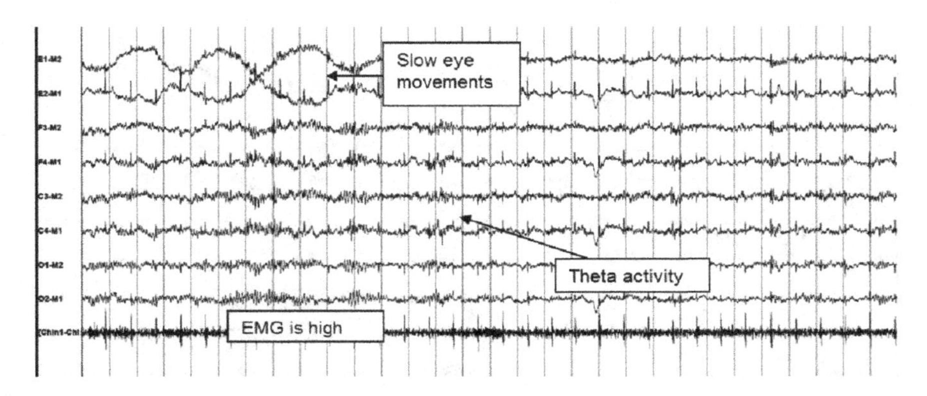

Figure 6.3 EEG, EOG, and EMG activity during Stage 1 sleep

this very light stage of sleep might not even remember that he/she was asleep. This is the stage we usually are in when fatigue leads us to "nod off" during a meeting, while watching TV, or even while driving or flying. When evidence of Stage 1 sleep is observed in someone who is supposed to be awake, it is often referred to as a *microsleep* or a *microlapse* (a short lapse into sleep). While relaxing at home, these short lapses are of little or no concern, but on the job, microsleeps can be very dangerous for three reasons: (1) they often occur involuntarily, even in people who are trying to stay awake; (2) as mentioned above, people who awaken from a microsleep often fail to realize that they ever fell asleep in the first place; and (3) trying to fly around while asleep is just plain scary!

Stage 2 Sleep

During a normal sleep period, Stage 1 lasts about five minutes before sleep progresses to a deeper stage – Stage 2 sleep. This stage of sleep is characterized by unique waveforms called k-complexes (intermittent very large EEG waves) and sleep spindles (high frequency EEG waves that "ride" on top of other waves in the overall pattern). Most researchers and clinicians believe the beginning of Stage 2 sleep rather than Stage 1 sleep is the true onset of sleep. The eyes generally are still, but they may move some. The muscles are somewhat relaxed, but still taut. The major change between Stage 1 and Stage 2 sleep is found in the EEG trace. An example of Stage 2 sleep is shown in Figure 6.4. Notice the sleep spindle and k-complex in the middle of the record.

Stages 3 and 4 Sleep

Stage 2 lasts about 10 to 20 minutes as sleep becomes deeper and the brain activity slows down even more. As time progresses, deeper stages of sleep

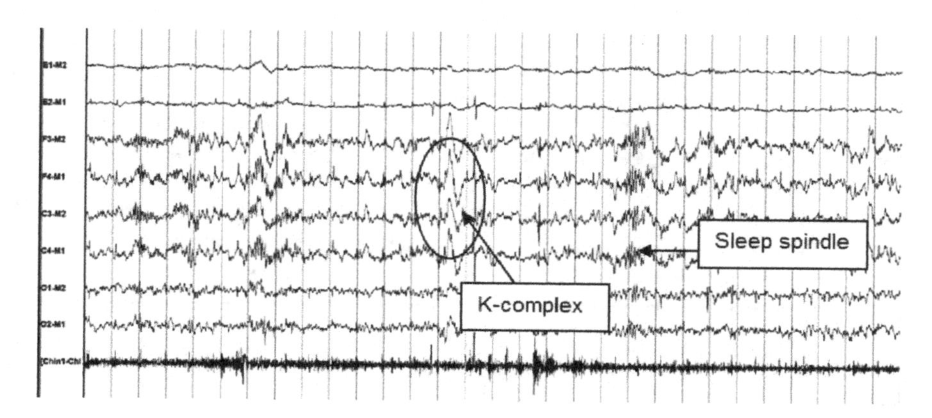

Figure 6.4 EEG, EOG, and EMG activity during Stage 2 sleep

are signaled by the appearance of slow, high amplitude *delta waves*, which have a frequency of between 0.5 and two cycles per second. These slow waveforms help to classify Stage 3 and Stage 4 sleep. Together, Stages 3 and 4 constitute *slow-wave sleep*, or delta sleep because the brain's patterns consist mostly of delta activity, as can be seen in Figure 6.5. When trying to awaken a person from slow-wave sleep, it may take a good deal more stimulation than what is required to awaken someone from Stage 1 or Stage 2 sleep. Furthermore, once someone is awakened from slow-wave sleep, they tend to be very groggy, and it may take several minutes until they are able to think straight. As mentioned earlier, this post-sleep grogginess is called sleep inertia, and one of the factors that is known to make sleep inertia severe is awakening from slow-wave sleep as opposed to awakening from Stage 1 or 2 sleep. At any rate, deep sleep is characterized by slow EEG activity and poor reactivity to the surrounding environment. Note that the eyes are still relatively inactive and the muscles are still taut, but a little more relaxed than they were in the lighter stages. Each of these epochs represents 30 seconds of the sleep recording.

Rapid Eye Movement (REM) Sleep

After spending about 30 minutes in slow-wave sleep, brain activity begins to become more active as it begins to move back into Stage 2 sleep for several minutes. The next progression of sleep is into REM sleep. During REM sleep, the eyes move rapidly from side to side, usually in bursts, as shown in the EOG trace in Figure 6.6. Of course, this is the characteristic that gives REM sleep its name. Besides the quick eye movements, the EEG activity of the brain becomes very fast and desynchronous. A novice looking at the brain activity in this type of sleep could mistake it as being indicative

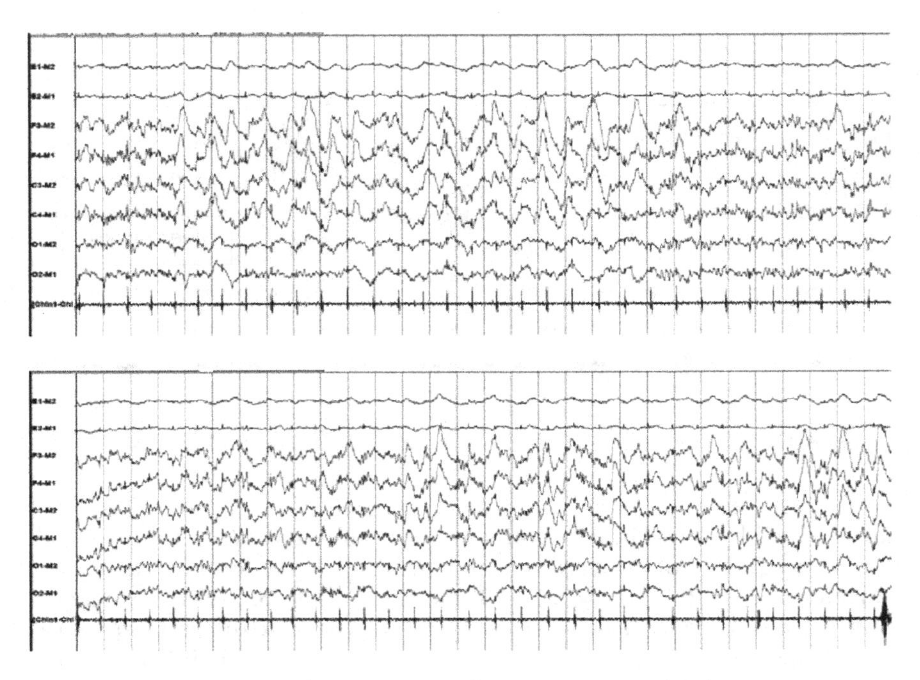

Figure 6.5 EEG, EOG, and EMG activity during Stages 3 and 4 sleep

of wakefulness, which is why some refer to REM sleep as paradoxical sleep. However, while the brain activity looks awake, the person is definitely asleep. Another factor that classifies this period of sleep as REM is the relative lack of muscle tone. The fact that the muscles are completely relaxed is an adaptive mechanism that the brain uses to prevent someone from moving around during this stage of sleep. Obviously, it would not be very adaptive from a survival standpoint if people were able to act out their dreams! Figure 6.6 shows the EEG, EOG, and EMG patterns during REM sleep.

There are several episodes of REM sleep during each night of sleep, and while the first REM period is very short (five to ten minutes before the brain reverts back to Stage 2 sleep), as the night progresses, the length of each REM period gets longer. As alluded to above, REM sleep is where most of our dreaming occurs. When scientists have awakened people from REM sleep, about 80 percent of the time they report dreaming, and this in not characteristic of awakenings from other stages of sleep. Since REM periods tend to occupy more of the sleep architecture right before awakening, many people wake from the night out of REM sleep, and many can vividly remember the dream that was occurring just before waking.

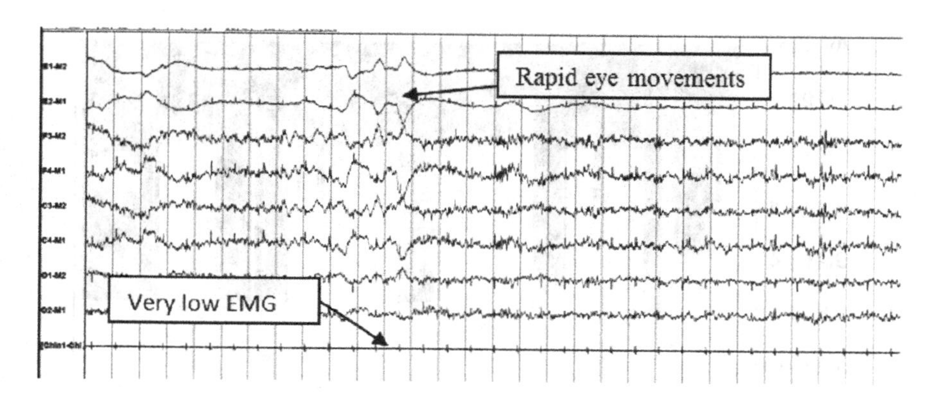

Figure 6.6 EEG, EOG, and EMG activity during REM sleep

The Distribution of Stages Throughout a Period of Sleep

The progression of sleep through the stages identified above occurs predictably during the night. Sleep begins in Stage 1 and then progresses to Stages 2, 3, and 4, prior to returning back to Stage 2, then to Stage REM. This cycle lasts approximately 90 minutes before it begins again. During the first half of the night, most of the slow-wave sleep occurs. During the second half of the night, most of the REM sleep occurs. Throughout the night, brief transitions to Stage 1 sleep and/or brief awakenings are often observed as well. These brief awakenings are common for adults, so they are considered normal. The progression of sleep stages over an eight-hour sleep period is shown in Figure 6.7.

Modifiers of Sleep Architecture

The progression of sleep stages is modified by age, time awake, time of day, environmental characteristics, sleep disorders, and a variety of other factors. The presence of these influences can affect the overall structure of sleep, and many will exert a noticeable subjective effect on the quality of sleep.

Sleep Deprivation

A common modifier of sleep structure is sleep deprivation. When job requirements entail sleep loss either because of extended duty periods or shift work, the sleep that occurs in the first rest opportunity often contains far more slow-wave activity throughout the night while REM sleep is pushed either much later in the sleep episode or sometimes even into the next night of sleep. The brain shows a preference for recovering slow-wave

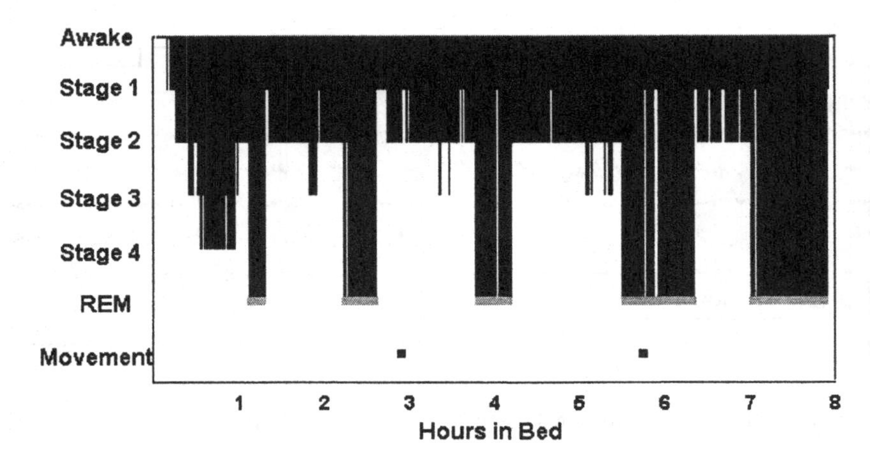

Figure 6.7 The progression of sleep stages throughout the sleep period

sleep first and REM sleep afterwards, and this leads us to believe that slow-wave sleep probably conveys a greater survival advantage than REM sleep. However, the exact roles of the different stages of sleep have not been precisely determined.

The extent of change in sleep characteristics will depend on how long the person has been awake before recovery sleep occurs. Figure 6.8 shows a hypnogram of the recovery sleep of one individual after he had been awake for 40 straight hours. Compare this to the more typical pattern observed earlier. Notice how quickly he entered Stage 4 sleep. Had the recovery sleep lasted longer than eight hours, more slow-wave sleep may have occurred toward the end of the night. This type of change in sleep architecture is important to note in situations where a sleep episode occurs immediately prior to the onset of work (such as when strategic napping is used as a workplace fatigue countermeasure). As noted earlier, sleep inertia will be greater when one is awakened from slow-wave sleep than when one is awakened from a lighter sleep stage. If a previously sleep-deprived pilot is taking a nap before resuming duties at the controls, it is important to remember where slow-wave sleep falls in the cycle so that extra time can be allotted for grogginess to dissipate before sitting down at the controls (note that this problem can be minimized by placing naps earlier in the sustained-wakefulness period, as will be discussed later).

The Timing of Sleep

The time at which one sleeps also can affect the structure of the sleep cycle. Day sleep usually has most of the REM activity in the first part of the sleep

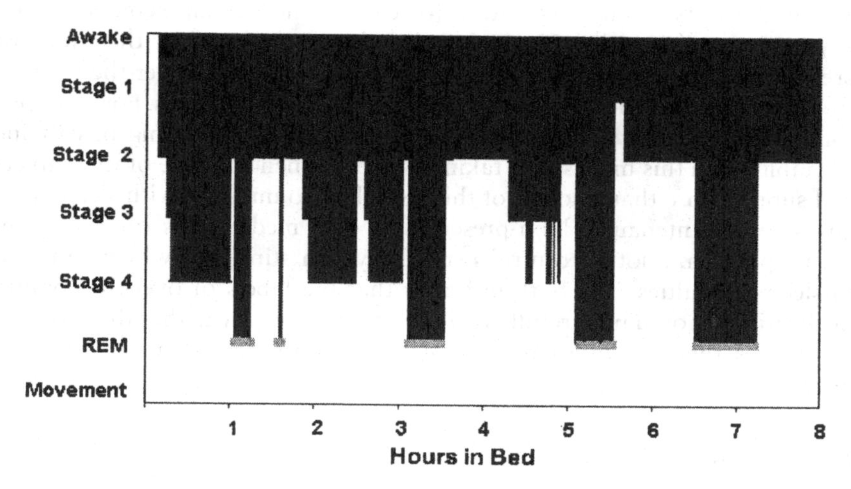

Figure 6.8 A hypnogram of recovery sleep after 40 hours awake

period, while the slow-wave sleep shifts to the second half of the sleep cycle. Shift workers who change from night sleep to day sleep may show reversed sleep pattern (REM first, slow-wave sleep second) on the first few days before their circadian cycle adjusts to the new schedule. Obviously, such a change can noticeably affect the restorative quality of sleep during the transitional period.

Age

Age also affects the sleep cycle. As infants, most of our sleep time is spent in REM sleep, but as we grow older, the amount of slow-wave sleep increases until we reach adulthood. As we enter our third decade of life, the amount of slow-wave sleep declines, until by the time we reach our 60s and 70s, almost all the slow-wave sleep is gone. Also, as we grow older, we lose that ability to "sleep like a rock" throughout the night as we begin to experience more of those annoying awakenings. Frequent nighttime awakenings are more common during our older years.

Medications

Medications also can change the sleep cycle. Although pilots do not need to worry too much about prescription drugs, it is worthwhile to note that many medications will inhibit either the amount of slow-wave sleep or REM sleep. Some medicines interfere with the ability to go to sleep or to stay asleep, while others lead to excessive sleepiness. A qualified physician/aviation medical examiner (AME) should definitely be consulted about the potential effects of any medication (or even an herbal supplement) before anything new is ingested. Not only is it necessary to consider the effects of prescription medications, but it is important to be wary of over-the-counter medications as well because these can affect sleep and alertness. For example, some over-the-counter pain/headache remedies contain 65 mg of caffeine per tablet, and this means that taking the recommended dose of two tablets will supply more than enough of this stimulant to interfere with sleep onset and sleep maintenance. Non-prescription cold medications often contain pseudophedrin, another central nervous system stimulant which can lead to sleep difficulties. The bottom line is that the labels of over-the-counter medications should be carefully reviewed to determine whether they contain substances that can interfere with sleep or substances that can promote drowsiness.

Environment

Environmental factors likewise can either improve or degrade sleep quality. Temperature, noise, and lighting obviously must fulfill certain criteria in order to ensure the best possible sleep. A setting that is too hot or too cold will produce enough discomfort to cause frequent awakenings. Although too much heat can create a miserable sleep setting, cold also can be quite disruptive since the body loses its ability to regulate temperature during REM sleep. Loud, intermittent noises are problematic because the brain is unable to habituate to these types of sounds. Travelers often experience sleep disturbances when staying overnight in a strange place because of even minor noises from hallway conversations, street traffic, or an air conditioner that cycles on and off throughout the night. Too much light also can present serious sleep problems since a bedroom that is not completely dark will permit one of the body's primary external cues (light) to encourage wakefulness at inopportune times.

Later on, several strategies will be discussed that are designed to promote the most restful possible sleep. At the point at which these are reviewed, it will become clear that many of the suggestions are designed to minimize the sort of sleep-disrupting influences that are within personal control. Paying attention to these suggestions represents the first step toward maximizing on-the-job alertness by minimizing homeostatic sleep pressure. However, before we get to this topic, it is important to recognize that some alertness difficulties are a function of medically recognized sleep disorders as opposed to more controllable factors. As the next chapter will show, these disorders can impair alertness because of their interference with the sleep cycle – either by causing frequent awakenings or by disrupting the phases of deep sleep.

Top Ten Points About Sleep Facts

- Getting seven to eight consolidated hours of sleep per night is the number one priority to maximize alertness.
- Sleep is a physiologically active process that can be categorized into defined stages.
- Stage 1 sleep is the transition from wakefulness into sleep.
- Stage 2 sleep is the first stage that counts toward restoring the body and the brain.
- Stages 3 and 4 sleep are most important for recovering from the fatigue from prior wakefulness.
- REM sleep is a lighter stage of sleep in which dreams occur.
- During REM sleep, the activity of the sleeping brain looks very similar to the activity of the awake brain.
- The brain cycles through the stages of sleep at 90-minute intervals throughout the night.
- There is more slow-wave sleep in the first part of the night and more REM sleep in the second.
- The sleep process is influenced by fatigue, bedtime, age, medications, and environment.

7 Sleep Disorders

As has already been stated, the average person needs about seven to eight hours of sleep during every 24-hour period throughout his or her adult life. Some people may need a little less and some people may need a little more, but seven to eight hours is the established, normal amount. Unfortunately, despite what most of us need to be at our best, it is a fact that many of us try to ignore this need, and as a result, most people do not get the amount of sleep that they require. Some limit their sleep because they would rather spend the time in social activities, recreational activities, or work; but other people do not get enough sleep due to an inability to sleep. They try to get the amount that they know they should, but something seems to thwart their good intentions. The people that fall into this latter category may be suffering from the effects of a medically recognized sleep disorder.

There are almost 100 sleep disorders recognized by the medical community. These disorders are classified into four major categories: (1) dysomnias – disorders initiating or maintaining sleep and disorders of excessive daytime sleepiness; (2) parasomnias – disorders that are characterized by problems that occur during sleep, but do not lead to insomnia or excessive sleepiness; (3) other sleep-disrupting disorders associated with medical or psychiatric problems; and (4) poorly defined "catch all" problems that contain sleep-related disorders that are too poorly understood to classify into anything else (but they have been proposed as possible sleep disorders). Of these sleep disorders, we will concern ourselves primarily with those that make up the first category, dysomnias. Sleep apnea syndrome is one of the most common of these, along with periodic limb movements in sleep (PLMS), narcolepsy, and insomnia.

An important step in treating any sleep disorder is the correct diagnosis. Proper diagnosis occurs after the patient has visited a sleep specialist for medical and family history review, has undergone a thorough physical examination, and has usually been studied during an overnight stay at a sleep laboratory. Once the sleep specialist determines that a sleep disorder is in fact present, and once the exact nature of the disorder is known, proper treatment can begin. A successful course of treatment can produce amazing improvements in both nighttime sleep and daytime alertness. In fact, the difference is often so pronounced that the patient feels like a different person as the symptoms associated with years of fatigue and discomfort suddenly disappear! This sort of change, coupled with the fact that several of these disorders can be successfully treated in a way that does not affect flying status, makes the pilot who has been living with such a problem wonder why he or she ever waited so long to see a sleep specialist in the first place.

A few of the most common sleep disorders along with their treatments are discussed briefly below. If more information is needed about these disorders or others, the web sites of the American Association of Sleep Medicine (www.aasm.org) or the NSF (www.sleepfoundation.org) should be consulted.

Sleep Apnea

A sleep apnea is defined as the cessation of breathing during sleep for at least ten seconds. The sleep of a normal person contains as many as five sleep apneas per hour (the number per hour is called the *apnea index*) without any residual effects on health or daytime alertness. However, when the apnea index increases to as much as 20 or more, then both health and daytime alertness suffer. Apnea events result from airway constrictions that are so tight that they do not allow enough oxygen for normal breathing to come through. These obstructions lead to snoring and gasping, but also to sleep disruption because the oxygen-deprived sleeper usually reacts to the lack of oxygen with a jerk that produces a shift into a lighter sleep stage (or a complete awakening). As these events occur throughout the night, sleep becomes fragmented and non-restorative. The severity of the symptoms is linked to the severity of the apnea. A person with severe sleep apnea can stop breathing over 100 times an hour, with episodes lasting as long as 60 seconds. A person with minor sleep apnea may stop breathing only ten times an hour with each episode lasting only ten seconds.

Symptoms

People with sleep apnea often do not fully wake up every time they experience an apnea event, so they may not complain of a "sleep problem." However, they sometimes will complain of restless sleep, and they usually complain of excessive daytime sleepiness, headaches in the morning, depressed mood or personality change, possibly impotence, inability to control the bladder, and/or a decline in mental performance. Their bed partners (or in really bad cases, their next-door neighbors) may complain of loud snoring which many times ends in a snort and awakening (or partial awakening) from sleep. Many people with apnea are overweight, and the incidence of apnea increases with age. Sleep apnea is seen more in men than in women. Untreated apnea that progressively worsens over time can lead to serious health problems such as high blood pressure, stroke, and increased risk of accidents due to increased sleepiness. Treatment of apnea can significantly improve all these associated problems.

Diagnosis

Identification/diagnosis of sleep apnea usually begins with complaints from the bed partner about loud snoring, and this prompts the sufferer to seek treatment. Proper diagnosis and treatment of sleep apnea occur after an overnight stay in a sleep laboratory where the physiological measures of brain activity, heart rate, respiration, limb and muscle movements, and oxygen saturation are monitored while the patient sleeps. Following documentation of the magnitude of the problem, treatment options can be investigated.

Treatment

Sleep apnea can be treated a number of ways, depending on the severity of the problem, the physical construction of the nose and throat, and the weight of the patient. The most successful treatment is through continuous positive airway pressure or CPAP. This treatment consists of a small mask that fits over the patient's nose and mouth and is linked to a small air pump. The pump sends air through the mask and acts as a splint to the airways, holding them open and preventing the airway collapse which leads to the inability of the patient to breathe properly. Prevention of the airway collapse alleviates snoring as well as the apneas, and this leads to less disrupted sleep during the night. As sleep becomes less fragmented, daytime sleepiness diminishes; the heart and brain have a continuous flow of oxygen, and the patient's mood and personality return to normal. Many times the patient loses weight from the increased activity that is secondary to the energy improvement resulting from better nighttime sleep.

Other treatments for sleep apnea include surgery on the upper airway that can be as minor as removal of tonsils and adenoids and/or straightening of the nasal areas. However, surgery may be as major as reconstruction of the jaw and tongue. Usually any surgical treatments occur as a series of surgeries, each of which is followed by a re-evaluation designed to determine the extent of the improvement after each process. The success of surgery depends on the reason for the airway obstruction; if the obstruction can be removed, the surgery is successful.

Another treatment for some apnea patients is weight loss. As might be expected, overweight people have additional fatty tissue surrounding the airways, and this tissue can obstruct proper breathing. In some situations, weight loss can resolve the problem entirely. Often when overweight sufferers of sleep apnea are placed on CPAP, they reap a secondary gain in terms of weight loss once their improved energy leads to increased physical activity. Many people can lose enough weight so that CPAP is no longer needed, or the pressure required to keep the airways open can be decreased, making the CPAP more comfortable.

Types of Apnea

There are three different types of sleep apnea. The most common is the *obstructive* sleep apnea described above. It derives its name from the fact that the breathing difficulties result from a physical airway obstruction. Another type of sleep apnea is *central* sleep apnea. This occurs when the signals from the brain do not stimulate the respiratory system, and the patient does not breathe for several seconds. The problems associated with this type apnea are the same as for obstructive apnea – poor sleep, excessive daytime sleepiness, and so on, and the treatment is usually either CPAP or medication. Sometimes central sleep apnea is associated with other medical disorders, and when these disorders are treated, many times the central apneas improve. The third type of sleep apnea is *mixed*, a combination of obstructive and central apnea. Treatment for this type of apnea usually focuses on the obstructive part of the apnea which, when treated, may lessen the central apneas as well. A *hypopnea* is a phenomenon that is related to apnea. This can occur when one has difficulty breathing during sleep due to a partial obstruction, but the airways do not completely close down as in an apnea. It is treated the same as obstructive sleep apnea.

Apnea and Flight Status

In the civilian aviation community, a pilot may still fly if diagnosed with sleep apnea as long as treatment is successful and compliance with treatment can be demonstrated. Usually, after diagnosis and treatment has begun, the pilot is brought into the sleep laboratory for a series of tests that measure alertness. The most important test is called the Maintenance of Wakefulness Test (MWT) because it provides an objective measure of how well a person can stay awake. If the pilot is able to "pass" this test, then he is considered fit for flight duty. However, the FAA, the Aviation Medical Examiner (AME), and/or the airline will probably require a record of compliance with CPAP treatment, and/or proof that the apnea has been successfully treated through surgery. Most CPAP machines now have the ability to track the hours of use, so compliance is easy to verify. A second overnight sleep test following surgery will show proof that the surgery was successful.

In military aviation, a pilot diagnosed with sleep apnea cannot continue his or her flying career if their apnea is treated with CPAP. However, successful treatment via surgery or weight loss is waiverable. The reason for loss of flight status if CPAP is required stems from the fact that the pilot cannot be deployed with a CPAP machine because this electrically powered device may not work in a field environment. Obviously, successful treatment with surgery or weight loss resolves the issue without complications.

Periodic Limb Movements in Sleep (PLMS)

Another sleep disorder that disrupts and fragments sleep is periodic limb movements in sleep or PLMS. This disorder can disturb sleep to the point that its restorative value is significantly degraded, leading to problems with daytime sleepiness.

Symptoms

This disorder involves the arms and/or legs (or both sets of limbs) jerking in a regular pattern during the night. The most common type of PLMS involves the legs. The severity can range from a mild toe movement to a full kick of the leg. Often the sufferer of PLMS is not aware of the limb movements during sleep, so the bed partner often reports the problem (when they tire of being kicked in the side all night long). The reason PLMS is considered a sleep disorder is that the movement of the limbs disrupts sleep. Each time a limb movement occurs, it can lead to a lightening of the sleep stage, or full awakening.

Diagnosis

In order to diagnose PLMS, the patient should stay overnight in a sleep laboratory while brain, muscle, heart, and respiration are monitored, just like the procedure required for diagnosing sleep apnea. Sensors attached to the arms and legs are used to identify and count the muscle movements associated with PLMS. Typically, there is a rhythmic limb movement that occurs every two to 30 seconds over a period of time.

Treatment

Sometimes sleep is not disrupted by these movements. In this case, no treatment is necessary. However, if there are disruptions in sleep architecture, a disorder is diagnosed, and medication is prescribed in an effort to reduce or eliminate the limb movements. The patient can try large doses of vitamin E, but this treatment only works in a small percentage of people. Other medications can be prescribed, usually a dopamine agonist or other medications that are prescribed for other types of movement disorders such as Parkinson's Disease (by the way, even though the treatments are sometimes the same for the two problems, there is no indication that people with PLMS are susceptible to Parkinson's Disease). Some physicians prescribe a benzodiazepine medication to treat PLMS. This medication may not stop the limb movements, but it can protect sleep enough that the movements are no longer disruptive.

A Variant of PLMS – Restless Leg Syndrome

A similar disorder that may occur with PLMS is called Restless Leg Syndrome (RLS). This problem actually occurs while a person is awake, but the discomfort, which intensifies in the evening, prevents or delays the onset of sleep. The symptoms of RLS vary from mild tingling in the legs to severe discomfort and pain that become pronounced when the person is sitting or lying down. To eliminate or reduce the unpleasant sensations, the sufferer usually needs to move the legs, and this adversely affects sleep. Many people with RLS may also have PLMS. The treatment for both disorders is the same.

Limb Movement Disorders and Flight Status

Unless pilots with this disorder are fortunate enough to respond to vitamin E therapy, some type of prescription medication will be necessary. Therefore, there may be problems maintaining flight status. The administration of benzodiazepines or dopamine agonist usually requires a grounding period after dosing due to the sedative characteristics of the drugs. Some of the other Parkinson's Disease medications do not have the sedative quality of the more common medications, but they may have some other side effects which are not compatible with flying (and not allowable under FAA or military regulations).

Narcolepsy

Narcolepsy is a sleep disorder whose main characteristic is excessive sleepiness. This is a fairly serious disorder that is often resistant to complete alleviation.

Symptoms

Narcolepsy is a disorder that usually begins during adolescence or young adulthood and is characterized by episodes of irresistible sleepiness, sometimes accompanied by muscle weakness (usually brought on by emotions – termed *cataplexy*), nighttime hallucinations, and disturbed nighttime sleep. Some or all of these symptoms may be present, from mild to severe levels.

Diagnosis

The diagnosis of narcolepsy requires an overnight stay at a sleep laboratory during which (once again) the brain, muscle, eyes, heart, and respiration are monitored. Since sufferers often seek treatment because their excessive daytime sleepiness is impairing their ability to function in day-to-day activities, other possible sources of their impaired alertness are checked first. Once the overnight stay permits disorders such as sleep apnea and PLMS to be ruled out, the patient is required to stay the next day for a series of four or five naps – a procedure called the Multiple Sleep Latency Test (MSLT). For this test, the patient lies down for 20 minutes every two hours and tries to fall asleep while the standard physiological parameters mentioned above are measured. The time it takes to fall asleep on each occasion is recorded. Once the patient is asleep, the brain is monitored for the type of sleep that occurs. If sleep occurs within 20 minutes across most of the naps, and if REM sleep occurs during at least two of these 20-minute naps, a diagnosis of narcolepsy is made and treatment options are discussed.

Treatment

Depending on the severity of the disorder, naps can sometimes be used to alleviate the excessive daytime sleepiness, but most of the time stimulant medication is required. If cataplexy (loss of muscle tone) occurs, medication can be prescribed for this as well. Usually all symptoms can be alleviated or reduced with proper medication.

Narcolepsy and Flight Status

A diagnosis of narcolepsy ends the flying career of a pilot. Even with medication, neither the FAA nor the military allows someone with narcolepsy to pilot an aircraft. The possibility of falling asleep or experiencing a cataplexy attack during flight is, of course, too dangerous. However, as you might have expected, the diagnosis and treatment of this disorder in anyone who is responsible for operating complex machinery (especially in the transportation industry or the military) is important to the safety of both the patient and others who are dependent on his/her actions.

Transient Insomnia

Insomnia is the inability to sleep during the night. If this inability occurs regularly (that is, for more than three weeks), it falls into the category of chronic insomnia. However, if the problem is short term, it is classified as transient insomnia.

Symptoms

Sleep problems consist either of difficulties going to sleep, remaining asleep, and/or an inability to return to sleep after awakening. Transient insomnia as well as chronic insomnia are often symptomatic of another problem such as stress, anxiety, apprehension, pain, new surroundings, and so on. In addition, short-term insomnia can be caused by environmental factors and circadian disruptions. People with insomnia usually seek treatment either because of their frustration over the inability to take advantage of sleep opportunities or because of the fact that their sleep problems are impairing alertness during waking hours.

Diagnosis and Treatment

The diagnosis of insomnia is made by acquiring a medical history and discussing life events and sleep patterns. Some of the reasons for insomnia may be quickly identified and thereby rectified once the specific circumstances under which the sleep problems occur are discovered. Of course, this often is not as easy as it may sound. As detailed below, there are many common possibilities that must be considered. At a minimum, stress, environment, and circadian factors should be explored when insomnia occurs in an otherwise normal, healthy individual. Anyone who is in a position of responsibility and who finds that their job often takes them away from home can easily relate to difficulties in all three of these areas. Once the source of the problem is clear, a specific treatment option can be identified and implemented.

Insomnia Due to Stress – Causation

Everyone lives with stress, at least to some extent, on a daily basis. Perhaps the source is an argument with a spouse or a friend, a disagreement with a coworker, financial problems, or even concerns about what the next day will bring. Sometimes, the source of stress can be a positive event such as a well-deserved promotion at work or finally buying that new house that has long been hoped for. Whatever the source, stressful events can temporarily disrupt sleep until either the source of the stress is removed or the individual has learned to deal adaptively with the stressful situation. If sleep suddenly becomes difficult for no apparent reason, perhaps a psychological stressor is the underlying source.

Insomnia Due to Stress – Treatment

To keep this temporary problem from becoming chronic, the individual should try to alleviate the source of the stress as early as possible so that the insomnia stays short term and does not develop into a chronic problem. Take stock of the current situation and implement one of many strategies designed to cope with psychological stress. These strategies are beyond the scope of this book, but a quick trip to the local bookstore is a great place to start. There is a plethora of advice available on how to overcome worry and transient anxiety, and some techniques are immediately effective. For instance, progressive relaxation exercises (systematically tensing and relaxing the body's muscle groups) have been proven very effective for dealing with temporary anxiety because these quick and easy, self-help strategies bring the body's natural relaxation response under more voluntary control. In addition, the act of focusing on producing a relaxation response may be all it takes to divert attention away from an anxiety-provoking situation toward a sleep-conducive state long enough to fall asleep. This may sound overly simplistic, but it is true that people can consciously identify and cope with a previously unrecognized problem once they focus energy in this direction. In the meantime, it would be worthwhile to engage in some type of aerobic exercise three to four hours before bedtime because this has been shown to improve the onset and quality of sleep later on. So, if psychological stress is producing transient sleeping problems, self-help techniques, relaxation exercises, aerobic exercise, or some other stress-management strategy may offer a solution.

Insomnia Due to Environmental Changes – Causation

A common reason for transient insomnia is environmental. When traveling to an unfamiliar environment, difficulty going to sleep may occur for a night or two until adjustment to the new environment is achieved. Also, the presence of noise during the night can interfere with sleep onset or cause awakenings followed by an inability to return to sleep. The presence of light in the room may prevent restful sleep, particularly if the light is intermittent. A room that is either too hot or too cold will interfere with sleep, as will a bed that is either too hard or too soft, a pillow that doesn't support the head and neck properly, or a sleep partner who is restless, snores, or otherwise disturbs the tranquility of the sleep setting. Deployed military pilots are especially susceptible to the adverse effects of a poor sleep environment. Sharing a tent with several other people, trying to sleep in a cot, in the desert, in the middle of the day, while listening to the sounds of aircraft departing and landing, trucks driving through the area, and people talking as they walk from one place to another can thwart even the best efforts to grab a few hours of sleep before the next duty cycle.

Insomnia Due to Environmental Changes – Treatment

Treatment of insomnia due to environmental factors usually involves alleviation of the noise, light, and so on. If the environment is temporary such as a business trip, then a hypnotic (sleep-promoting medication) may be prescribed to induce sleep during the short time involved in the trip, but for pilots, this generally is not an option. Problems sleeping in an unfamiliar sleep setting are common for people who travel a great deal. Sometimes the environment itself actually seems comfortable and conducive to restful sleep, but there is just something about trying to sleep "in a strange bed" that causes trouble, at least for one or two nights. People who have transient insomnia associated with unfamiliar environments need not feel that they are unusual. This is why sleep researchers usually require patients or study participants to sleep in the sleep laboratory for one night (referred to as an "adaptation night") before they actually collect the sleep data that will be used to diagnose a sleep problem. Humans are creatures of habit, and when the routine is disrupted it takes time to readjust. The speed of this adjustment may be slightly improved by taking something familiar from home to the new sleep quarters. Examples are a comfortable pillow, a family picture, or even minor convenience items that might not be available in a typical hotel room or temporary military quarters. Improving the environment in military field settings is extremely difficult. However, the use of sleep masks can block light, foam ear plugs can attenuate noise, and cooperative agreements with tent mates can minimize the disruptive effects of conversation and other activities in the sleep quarters.

Insomnia Due to Circadian Disruptions – Causation

Another source of short-term insomnia is related to the sleep schedule. As was discussed in a previous chapter, shift workers have a very difficult time maintaining sleep due to the time at which they have opportunities to sleep. Work/rest cycles that vary frequently are often associated with sleep difficulties because the body cannot establish a consistent routine. In addition, traveling across multiple time zones can produce sleep difficulties similar to those experienced by the shift worker. Long-haul pilots are constantly confronted with disagreements between their body's internal rhythms and the environmental cues present at their destination. Military pilots likewise must face these problems when they are deployed across multiple time zones, especially when traveling in an easterly direction. A pilot who is working a normal daytime schedule in the US may be in the habit of waking up early in the morning and going to sleep at 22:00 or 23:00 each night. As noted earlier, such a schedule means that the time span from approximately 17:00–21:00 will be associated with a high degree of alertness. This is fine while staying in the US. However, when this pilot travels from Atlanta to Frankfurt, he has crossed six time zones in less than a day, and now, when he attempts to stick to his 23:00 bedtime in Germany, he is actually trying to sleep at 17:00 according to his body's internal clock. Needless to say, it is very difficult to fall asleep when the body thinks it is six hours before bedtime, especially when this is a time associated with high circadian alertness. And if the problem of getting to sleep were not enough, the problem of getting the body to awaken at a time that is six hours before the usual body clock wake up is fairly unpleasant as well, especially when most of the night was spent staring at the ceiling in frustration rather than sleep.

Insomnia Due to Circadian Disruptions – Treatment

Insomnia from circadian disruption is difficult to manage, particularly in aviation settings. Our bodies cannot adjust quickly to changes in schedules, and sleep is generally one of the casualties of a new schedule. In some situations, hypnotics (sleeping pills) may be prescribed for the short term. However, implementation of behavioral countermeasures also represents a successful alternative. The application of countermeasures depends in part on the length of time the person will be in the new time zone or on the new work schedule. Problems and solutions for both shift lag and jet lag are discussed further in Chapter 11.

Transient Insomnia and Flight Status

There are no specific criteria under which insomnia will impact the ability to secure and maintain flight status. Many cases are relatively mild and, because of this, they have an inconsequential effect on routine alertness levels. Also, even when a transient insomnia is more severe, the short duration of the sleep difficulty is unlikely to result in any type of chronic impairment. However, it should be remembered that any degree of sleep loss translates into some decrement in alertness later on. Because of this, countermeasures for both shift lag and jet lag (soon to be discussed) should be implemented.

Chronic Insomnia

As noted earlier, a chronic insomnia is a sleep disturbance that lasts for longer than three weeks. As is the case with transient insomnia, the difficulty may be in the initiation or the maintenance of sleep.

Symptoms

The primary symptom is the relatively long-term problem of going to sleep once in bed or remaining asleep throughout the sleep period. The insufficient sleep that results from these disturbances often causes alertness decrements on the job and elsewhere.

Diagnosis and Treatment

Chronic insomnia is much more difficult to treat due to the various reasons for its occurrence. Sometimes the person has developed habits that are not conducive to sleep or he or she may be taking a medication that interferes with sleep. If the problem is as simple as this, the insomnia can be treated fairly successfully. However, if the insomnia results from a medical disorder or another sleep disorder, a more complex solution is needed. Pain can interfere with sleep as much or more than sleep apnea or PLMS. Mitigation of the discomfort or treatment of the sleep disorder will usually alleviate the insomnia. Regardless of the source, chronic insomnia is a serious condition due to the adverse effects on both health and safety. As such, it should be discussed with a physician – preferably a Board-Certified Sleep Specialist.

Chronic Insomnia and Flight Status

Pilots who experience chronic insomnia may be successfully treated with the behavioral interventions that will be discussed later on. In addition, cases of chronic insomnia that are the result of environmental factors can often be rectified by modifying the characteristics of the sleep setting. Cases that are due to circadian factors can also be corrected without medication, although it is difficult to eliminate the adverse effects of constant schedule changes in transcontinental pilots or frequently deployed military aviators. A variety of tips for fixing sleep problems with "non-medical" interventions will be detailed in subsequent chapters. The bottom line is that chronic sleep disturbances can often be corrected without affecting fitness for full flying duty. However, the first step is to seek professional help from a sleep specialist. The possibility of medically recognized sleep disorders must be seriously considered and ruled out to ensure that optimal sleep on the ground supports peak performance in the air.

Top Ten Points About Sleep Disorders

- Medically recognized sleep disorders can cause severe fatigue due to sleep disruption.
- An important step in treating any sleep disorder is the correct diagnosis.
- Apnea is a breathing disorder that produces numerous awakenings throughout the night.
- Limb movement disorders impair sleep because of unconscious muscle jerks or twitches.
- RLS delays sleep onset due to pain that can only be alleviated by moving around.
- Narcolepsy causes severe uncontrollable sleepiness (and other problems) even in the middle of the day.
- Temporary sleep problems (one to six nights) often result from stress, environmental, or circadian factors.
- Long-term sleep problems (three or more weeks) result from a number of factors including sleep disorders and bad sleep habits.
- Persistent sleep problems should be immediately brought to the attention of a physician.
- Many sleep-related problems can be corrected without adversely affecting flight status.

8 Other Factors Associated with Sleep Difficulties

In the absence of an identifiable sleep disorder, there are other circumstances that can lead to problematic levels of aviator fatigue. Many of these, such as long work hours and non-standard work schedules, are often not within the control of the individual. Someone has to pull the night shift in 24/7 commercial operations and pilots are often faced with travel across multiple time zones and other duty requirements that make it difficult to attain the amount of restful sleep necessary to be fully alert. However, there are some factors that often are, to some degree, within the individual's control. These include some transient insomnias discussed in the previous chapter as well as the problem of intentional sleep restriction and the problem of poor sleep habits. These latter two causes of insufficient sleep are important because they can create a serious and chronic increase in fatigue levels on the job, and they can degrade general well-being and quality of life. Fortunately, both can be completely corrected by the individual experiencing the problem without the aid of professional intervention and without the fear of an untoward impact of flight status.

Intentional Sleep Restriction

By now the importance of sleep in the fatigue equation is obvious. From personal experience, most people can identify with the lack of concentration, irritability, and lethargy that stem from insufficient sleep. However, despite this knowledge, many adults in the industrialized world continue to treat sleep as an option rather than as a requirement. This is an important point in light of the fact that many on-the-job problems with inadequate alertness result from simply going to bed too late at night and getting up too early in the morning. With work requirements, family obligations, shopping, recreation, the internet, and other activities, the temptation to squeeze an extra hour into each day by stealing an hour of sleep from every night is strong. The NSF estimates that in the last century, Americans have decreased their sleep time by approximately 20 percent while adding about one month each year to their commuting and working time. The average adult now sleeps less than seven hours a day despite the fact that sleep specialists recommend seven to eight hours for optimal functioning and well-being.

How important is a sleep reduction of one to two hours every night? Thanks to recent research efforts, the answer to this is fairly clear. Not long ago, it was thought that sleeping only five hours per night would not create substantial problems for most people as long as their sleep restriction was limited to a few days. But an investigation by researchers at the University of Pennsylvania Medical School found that such partial sleep deprivation quickly degrades the type of vigilance needed in aviation and other "high-attention" tasks. Figure 8.1 shows what happened to the attention levels of three groups of research participants who were exposed to varying levels of sleep. Before the sleep-restriction intervention, there were two baseline days on which everyone got eight hours of time in bed per night. Then, only one of the groups was allowed to continue their eight-hour schedule while the remaining two groups underwent partial sleep deprivation over 14 days. Note how rapidly performance began to deteriorate in the group that was restricted to only four hours per night. However, this was not the only group that suffered. As the study progressed, the six-hour time-in-bed group also lost their ability to maintain baseline levels of performance. Only

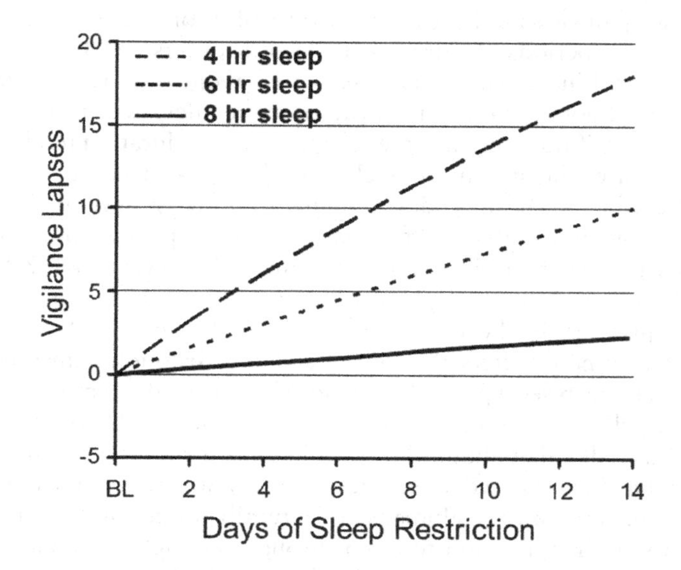

Figure 8.1 Chronic sleep restriction degrades performance

Source: Adapted with permission from Van Dongen, H.P.A., Maislin, G., Mullington, J.M., and Dinges, D.F. (2003), 'The Cumulative Cost of Additional Wakefulness: Dose-Response Effects on Neurobehavioral Functions and Sleep Physiology From Chronic Sleep Restriction and Total Sleep,' *Sleep*, Vol. 26(2), pp. 117–126.

the volunteers who were allowed eight hours in bed every day were consistently able to monitor and respond to the task at hand (results from other tests were similar). So, the bottom line is that while six hours of sleep is better than four, it still is not enough to maintain performance at optimal levels over a period of time.

In the study cited above, the sleep restriction was not "intentional" because the volunteers were assigned to a certain sleep schedule for the sake of the research. However, in the real world, people often purposely restrict their sleep simply as a matter of choice. They make a conscious decision to rob themselves of the precious commodity of adequate daily slumber. As a result, they unwittingly set themselves up for impaired daytime alertness, poor mental concentration, depressed or irritable mood, degraded immune-system functioning, and problems relating effectively to coworkers, friends, and family members. The take-home message here is that sleep must be prioritized along with other important daily activities or the resulting sleep restriction will quickly affect performance and quality of life.

Poor Sleep Habits

Up to this point, the reasons for fatigue-producing sleep deprivation have included sleep disorders, short-term episodes of insomnia, and intentionally restricted sleep periods. As discussed, sleep disorders can be treated at a sleep clinic, and intentional sleep loss can be avoided by reprioritizing the daily routine. Short-term insomnia can often be alleviated by a few simple tricks, but even if this were not the case, their short duration tends to make them less of a problem than more chronic sleep disorders. But what about the people who are healthy, don't often travel across several time zones, usually have a comfortable and familiar place to sleep, and who make every effort to set aside the right amount of time in bed every night? Many of these folks still find that they *just can't go to sleep*. If they are experiencing problems, just imagine how the source of their difficulties affects the flight crew that routinely suffers through circadian disruptions, uncomfortable or unfamiliar sleep settings, and unavoidably restricted sleep opportunities!

What could cause sleep problems for all of these people? As it turns out, even healthy folks that should be "sleeping like a rock" every night sometimes become their own worst enemies by the choices they make (either consciously or unconsciously). Humans are naturally programmed to fall asleep easily at the right times and to sleep through the night without difficulty. However, this natural tendency can be thwarted by lifestyle factors, dietary habits, and other practices that are simply inconsistent with good sleep.

We are Creatures of Habit

People are creatures of habit like everyone else in the animal kingdom. Next time one of those circus acts in which the animal jumps through the hoops, crosses the bridge, rolls the ball, and climbs the ladder leads us to chuckle about our superiority in the scheme of life, perhaps we should consider our own daily routines. How much of daily life is filled with real spontaneity and how much of it is filled with grinding routines that vary little from week to week? We may think we are not on a behavioral autopilot for large parts of each day, but close study will likely reveal that the opposite is actually true. Anyone who has ever tried to drive a car in one of those countries where everyone drives on "the wrong side of the road" can attest to the fact that old habits are hard to break. Most of the time, we perform routine tasks in exactly the same sequence day after day. Our performance quickly becomes so automatic that we do not even think about what we are doing. In part, this is because of a learning phenomenon called *chaining*.

Behavioral Chains

Most complex behaviors consist of a series of simple behaviors that are "chained" together into a uniform sequence. The getting-ready-for-work chain consists of turning off the alarm clock, climbing out of bed, going to the bathroom, taking a shower, drying off, combing hair, brushing teeth, getting clothes out of the closet, and so on. This chain can be broken into a series of smaller chains that consist of sequences of smaller individual tasks. Just think of all of the little things that make up the task of getting dressed! The interesting thing about behavior chains is that specific individual components of the chain come to act as cues for other behaviors in the chain. In fact, once a chain is established, the occurrence of one behavior actually increases the probability of the subsequent behavior in the sequence. This explains why certain activities are performed in exactly the same sequence every time. The individual components become connected so strongly that they occur together automatically without any higher-level thought. In addition, behaviors that are not very well connected are eventually automatically eliminated from the chain.

The "Up Side" and the "Down Side" of Chains

For the most part, these routines or habits are great because they free up cognitive resources for more demanding activities. Most people can get ready for work while thinking about what they will be doing several hours later in the day rather than thinking about tying shoes and buttoning shirts. In addition, well-practiced sequences of behaviors often reduce the probability of errors. So, habits (chains) can be helpful. However, habits can also create problems, and sometimes these problems interfere with sleep and result in operator fatigue. For instance, habits we do while in bed, like watching TV or paying bills or talking on the phone or arguing with a spouse, will eventually become strongly associated with being in the bedroom, while the habit of falling asleep will become less likely. Or perhaps a maladaptive chain will inadvertently be formed that dictates that sleep onset occurs only after the TV show, the paperwork, and the phone calls have been completed. Either way, sleep suffers because the "sleep chamber" has become a cue for all sorts of behaviors that are incompatible with sleep! Other habits can likewise lead to sleep disruption even before bedtime rolls around. Examples include the habit of drinking coffee after dinner or smoking a cigarette right before bedtime. Also, the practice of sleeping late on weekends or days off is common. Many people find that as they grow older (and as their sleep becomes more fragile) these routines suddenly begin to create insomnia problems that did not exist a few years earlier. If you developed bad sleep habits, think about the bad "chains" that got you into this mess, and commit yourself to forming good chains to correct the problem. All it takes is practicing the right sequence until it becomes automatic.

All By Themselves, Bad Habits Can Lead to Chronic Insomnia

Poor sleep habits ultimately can result in chronic insomnia that lasts for years and becomes so frustrating that the sufferer actually begins to dread going to bed at night. Once in bed, chronic insomniacs often lie awake for hours worrying about how they will not be able to function the next day because of their inability to sleep. The disconcerting fact is that many people unwittingly create this problem for themselves because they simply do not stop to think about the ramifications of their actions. In fact, some bad habits seem to be great ideas at the start. Doing an extra hour of paperwork every night has to be a good thing, right? Adding a few more minutes to enjoy some relaxing entertainment from TV at first seems like a great idea. Getting a head start on tomorrow's emails before going to bed tonight should ease the stress of the upcoming day. Unfortunately, these good ideas can backfire for some people. A perfect example of this is the use of technology (TV, cell phones, computers) right before bedtime. A recent survey revealed that 90 percent of Americans use some type of technology (most frequently TV) within an hour of bedtime, and that using technology in this way is associated with difficulties falling asleep and staying asleep. Mobile phones were viewed as particularly problematic. They are great for staying in touch while on the go, but they have no place in the bedroom! TV can be wonderful entertainment, but when watched too close to bedtime it can be a sleep disrupter. Computers have become an essential part of daily life, but the wavelength of light their screens emit can have a physiologically based alerting effect that is not compatible with sleep. Bottom line: Sleep and technology don't mix!! Needless to say, with all of the other fatigue-inducing problems that pilots face on a daily basis, anything that impairs restorative sleep should be avoided like the plague. In the upcoming discussion of operational fatigue countermeasures, one of the topics will center on techniques designed to maximize the benefits of available sleep opportunities. The development of good sleep habits is one of many important ways to improve performance, alertness, and well-being at home and at work.

Top Ten Points About Factors Associated with Sleep Difficulties

- Fatigue-causing sleep problems often are within the control of the individual.
- Many people sacrifice needed sleep for the sake of recreational, family, or work activities.
- Such reductions in the amount of nightly sleep (as little as one hour) can increase fatigue levels.
- If you are intentionally sacrificing sleep, stop! If you're trying to sleep but can't, check your sleep habits.
- Chronic insomnia can be a self-inflicted result of poor sleep habits.
- There are many lifestyle, dietary, and other habits that can sabotage restful sleep.
- Engaging in non-sleep-conducive behaviors in the bedroom can connect the bedroom with staying awake instead of going to sleep.
- Using technology right before bedtime can seriously impact sleep quantity and sleep quality.
- The good news is that bad sleep habits can be replaced with good ones.
- The sleep that comes with optimal "sleep hygiene" is essential for optimal on-the-job performance.

Part III

Countermeasures for Fatigue

9 The Need for Fatigue Countermeasures

Given the causes of fatigue and the disastrous consequences that tiredness in the cockpit sometimes produce, there is little doubt that combating this insidious and pervasive threat to aviation safety deserves high priority. Obviously, the best fatigue countermeasures are preventative strategies that are designed to ensure adequate sleep and optimal circadian adaptation before reporting for duty. However, as was previously discussed, both civil and military aviation operations often are not conducive to promoting the daily sleep and rest that is essential for the best possible performance.

Indicators of Increased Fatigue Risk

Many aviators find that their work schedules contain virtually every single warning indicator proposed by Dr Laurence Hartley, an expert in the effects of fatigue in transportation operations. Note the following characteristics of schedules and behaviors that are indicative of a fatigue-related threat to safety and effectiveness:

- often fly at the sleepy time of day (pre-dawn, early morning);
- work 12–14 hours in a day;
- often have nine or fewer hours off duty per day;
- have little or no time off in the previous seven days;
- frequently fly long, continuous flights;
- work inflexible schedules that don't allow short rest breaks or naps;
- are involved in a lot of staff or administrative duties in addition to flight duties;
- often fly after less than seven hours sleep;
- constantly work irregular schedules;
- chronically use alcohol to promote sleep;
- chronically rely upon caffeine, energy drinks, or herbal substances to stay awake;
- suffer from problems indicative of a sleep disorder.

Anyone who can readily relate to most or all of these risk factors is in danger of making serious errors of omission or commission, or even falling asleep at the controls because of high levels of fatigue. The reasons have already been discussed, but are worth a quick recap.

Many Schedules Increase Susceptibility to Circadian-Impaired Alertness

Pilots who work in the pre-dawn or early-morning hours, who constantly staff irregular shifts, and/or who repeatedly cross multiple time zones are battling their circadian rhythms by working at times when the body is programmed to sleep. Remember the circadian component of sleep and alertness? Furthermore, the requirement to rapidly adjust work/rest schedules from day, to evening, to night shifts, or during transoceanic routes (or military deployments) creates a persistent state of circadian desynchronosis (disruptions to the body's clock) that contributes to on-the-job problems with alertness, off-duty problems initiating and maintaining sleep, and a host of uncomfortable symptoms such as physical fatigue, gastrointestinal upset, poor mood, and so on.

Scheduling Factors Can Increase the Homeostatic Drive for Sleep

Aviators who work long hours every day, fly lengthy continuous routes, and have only minimal time off are clearly fighting the homeostatic drive for sleep. As noted earlier, the greater this homeostatic drive becomes, the more difficult it is to remain awake. When fatigued individuals have other responsibilities that prevent them from sleeping enough every day and taking some time off every week, they are building a cumulative sleep debt that can seriously impair overall alertness, performance, and well-being.

The Work Environment Can Exacerbate Underlying Fatigue

Add environmental/contextual factors to the presence of high circadian and homeostatic drives for sleep, and it becomes easy to see how performance and safety are at risk. Basic physiological fatigue is difficult enough to overcome even in the best of circumstances, but the typical flight environment – quiet, often dimly lit, highly automated, and virtually devoid of opportunities for physical activity – makes the situations much worse.

Inadequate Implementation of Fatigue Remedies Impedes Alertness Management

Pilots make every effort to endure the fatigue that stems from inadequate daily sleep and work schedules that are highly demanding. However, the limited availability of a range of valid fatigue countermeasures, the improper utilization of select anti-fatigue or circadian re-entrainment strategies, and/or the reliance on ineffective or inappropriate sleep-enhancement techniques leave some flight crews wondering how they will ever continue to meet the demands of their profession. Realistically, what is to be done?

There are Realistic Solutions to the Problem of Fatigue

Aviation operations rarely provide the predictable nine-to-five daytime work routines enjoyed by many workers, and this makes the straightforward implementation of several standard fatigue countermeasures difficult. Nonetheless, a variety of effective anti-fatigue strategies can be tailored to fit. When properly implemented, the use of one or more of these proven strategies will help to control the homeostatic, circadian, and other factors that threaten flight safety and personal well-being. However, before discussing specific recommendations, it is important to note that there are no simple, universal solutions that will be equally effective for every individual or every situation. As Dr Mark Rosekind (1994), former NASA scientist, founder of Alertness Solutions, 40th Member of the NTSB, and 15th Administrator of the National Highway Traffic Safety Administration (NHTSA) points out: (1) there are wide variations in workplace demands; (2) there are considerable differences in how individuals respond to these demands; and (3) there are changes in individual responses to both job factors and the specific countermeasures that occur over time. Thus, what works for a regional air carrier might not be appropriate for military bomber operations simply because the basic jobs are different. What helps a young Apache helicopter pilot might not be effective for one of his comrades simply because no two people are exactly alike. Even though both pilots are the same age and are flying the same missions in the same aircraft, variations in their physiological and psychological makeups will likely make them differentially responsive to their jobs and to the effects of various coping strategies. Likewise, the best fatigue countermeasure for a 45-year-old airline pilot may not do the trick for this same pilot at age 55 because of natural age-related changes in sleep architecture and/or circadian rhythms.

Alertness-Management Strategies Must Be Tailored

Jobs are different, people are different, and both tend to change with the passage of time. Because of this, one or two specific strategies will not provide the comprehensive coverage that is necessary. Unfortunately there are no magic formulas for deciding who will respond to any particular fatigue remedy or combination of remedies. Rather, more general advice is essential, and once it is given, each individual, crew, or unit must use some degree of trial and error to see what works. Most people do not buy clothes until after trying them on, and very few people buy a car until after the test drive. In a similar fashion, promising anti-fatigue strategies should be tested out for a reasonable period of time before any final conclusions are drawn. Furthermore, some situations may call for the implementation of two or three fatigue countermeasures to obtain the desired effect, and some individuals may find they need more than a single technique to promote optimal on-the-job alertness and the highest quality of off-duty sleep. Combining two or more proven strategies can significantly enhance the effectiveness of any alertness-management program, especially when day-to-day variations in job demands call for a more comprehensive approach.

The following chapters present an overview of the most well-known strategies for avoiding or managing fatigue on the job. Many of these focus on directly modifying the homeostatic or circadian components of sleep/alertness discussed earlier. This is the case with the preventative measure of ensuring adequate off-duty sleep, as well as the more operationally oriented strategies of proper shift scheduling, bright-light/hormonal manipulations, and strategic napping. However, there are other approaches that will be discussed as well, including the possible use of drugs in certain types of flight operations. The effects of caffeine are discussed since this is an option available to everyone. Finally, information on rest breaks, postural manipulations, exercise, environmental factors, and physical fitness will be reviewed either to explain the usefulness of these approaches or to point out that the effectiveness of a few of these strategies is limited or unsubstantiated.

Top Ten Point About the Need for Fatigue Countermeasures

- Controlling the insidious and pervasive effects of fatigue on the flight deck is essential to air safety.
- This is often difficult in civilian and military aviation for several reasons.
- Late-night and early-morning flights can severely degrade alertness.
- Short layovers and little "off time" contribute to an accumulation of fatigue.
- Irregular schedules disrupt the body's clock.
- Administrative duties and paperwork exacerbate the fatigue from long flight hours.
- The use of alcohol as a sleep aid and caffeine or other substances as a sleep substitute often makes things worse.
- The lack of systematic alertness-management strategies is a major contributor to the overall fatigue problem.
- However, there are scientifically valid techniques to combat aircrew fatigue.
- Specific anti-fatigue strategies can be tailored for individual pilots and their work situations.

10 Sleep Optimization

The First-line Fatigue Countermeasure

No matter what schedules are required and what conditions are present in the work environment, the most effective strategy for minimizing fatigue on the job is, of course, to ensure adequate sleep prior to the duty period. As noted earlier, insufficient sleep is an insidious threat to safety, performance, and personal well-being, and a substantial amount of research has shown that sleep restriction degrades performance in a dose-related fashion. In other words, the greater the amount of sleep deprivation or restriction, the more the loss of performance, mental clarity, judgment, and mood will be. The impact of chronic sleep loss is most noticeable in sedentary situations such as while watching television, reading, riding in a car, attending meetings, driving, flying during the cruise portion, and while performing routine work, but even in more active circumstances, the insidious effects of sleep loss are working to undermine performance and safety.

Will Sleep Loss Really Affect Pilot Performance?

How much sleep is necessary to prevent such decrements? As mentioned earlier, recent work has shown that the performance and alertness of most people deteriorate fairly quickly after restricting sleep to less than seven hours per night. It also has been established that reducing the amount of sleep to only four hours or less rapidly leads to serious uncontrolled "sleep attacks" that often occur without the knowledge of the victim. In about half of all cases, sleepy people who inadvertently doze off while they are trying to stay awake never even know they have fallen asleep until after they have been unconscious for a full minute! This may not sound like a big deal, but remember that someone who falls asleep for only five seconds while driving at highway speed will travel the length of an entire football field before they wake up again. That is a long way considering that those annoying obstacles like telephone poles, bridge rails, and oncoming cars are only 10 to 20 feet from the front of the vehicle. Just imagine how far a pilot on final approach at 150–200 knots will travel in the same amount of time. And as we have said before, brief sleep events of five or more seconds have actually

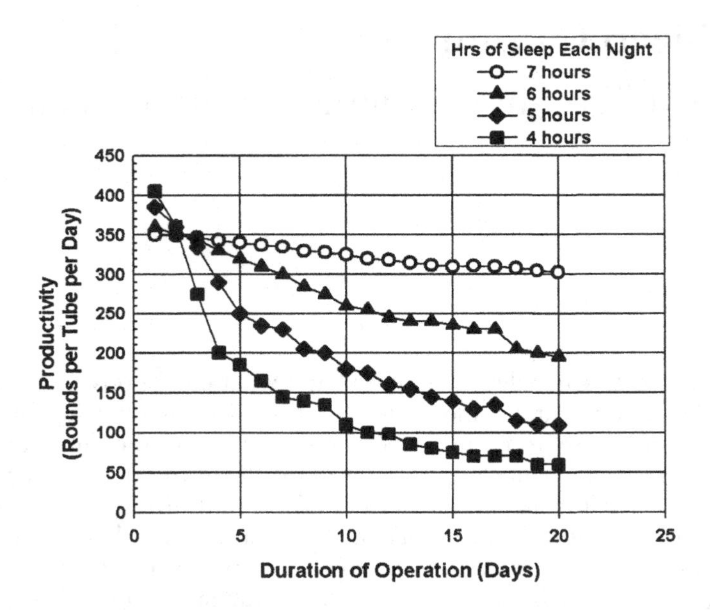

Figure 10.1 Performance rapidly falls off when sleep is restricted to less than eight hours per night

Source: Reprinted with permission from Belenky, G. (1997), 'Sustaining Performance during Continuous Operations: The US Army's Sleep Management System,' *Managing Fatigue in Transportation*, International Conference Proceedings, pp. 95–103.

been documented even in the last 90 minutes of flight – during descent and landing! The greater the sleep debt, the more frequently the problem was observed.

Although many personnel feel they can "gut it out" and adjust to chronic sleep restriction, there is no objective proof that this is the case. Studies have shown that people do not adapt to sleeping less than they need even after days of trying. Look at the data collected by the scientists at the Walter Reed Army Institute of Research. Figure 10.1 clearly shows that while people who sacrifice their sleep may initially be more productive at straightforward tasks (because they are continuing to perform while others are asleep), any benefits of sleep loss rapidly dissipate as fatigue takes over.

How Do I Know How Much Sleep I Need?

Determining individual sleep requirements is not easy since there is wide variability in sleep needs. Individual requirements range from about four to ten hours. This means that there are some people who are lucky enough to have a low physiological sleep requirement, and because of this, they can function effectively with only five or six hours of sleep per night. However, these people are exceptions to the rule, and they are not representative of most of the human population, which *needs at least seven full hours of sleep per day* to sustain adequate alertness. Unless you know for a fact that seven hours is more than you need, the safe approach is to make sure you get a minimum of seven full hours each day even if work demands mean you have to split this requirement into more than one consolidated period. In fact, the American Academy of Sleep Medicine and the Sleep Research Society in their 2015 Consensus Statement say that adults should obtain seven or more hours per day on a regular basis for good health and performance.

Two Ways to Calculate Your Individual Sleep Need

If you want to determine your own sleep need with a reasonable degree of certainty, try studying your sleep requisite during your next vacation. First, plan on putting the alarm clock away so that you can just wake up naturally every morning. Second, allow about two to three days to overcome any existing sleep debt, and get rid of daily schedules that would create a false bedtime/wakeup time. Then, for the next three or four days, write down what time you go to bed at night and what time you naturally wake up in the morning. Average the sleep hours across this period of time, and you will develop a good idea of your basic physiological sleep requirement. This is the amount of sleep that is necessary for optimal alertness, performance, and physical and mental well-being. Prioritize your life to make sure you get this amount. You will be amazed at the difference it will make at home, at the gym, on the job, and everywhere else!

A second way to estimate your sleep need is to systematically change the amount of sleep you get from one week to the next. If you normally get six hours a night, try increasing the amount to seven hours a night for one week. At the end of each day during this week, take a moment to reflect on whether you felt more alert as a function of the increase. Next week, change the sleep amount to eight hours per night, and once again, reflect on any changes in mood or alertness that result. Then, if you are not sure whether the change has made any difference, try reverting to the six-hour schedule for a couple of days before returning to the eight-hour sleep schedule. This technique of calculating your sleep requirement is not as accurate as the one mentioned earlier, but it has the advantage of being a feasible strategy even during normal work periods (as opposed to having to wait until the next vacation). If you make a dedicated effort to accurately monitor the changes in your feelings from one systematic schedule modification to another, this technique can be successful.

And remember, while experimenting with different sleep durations; ignore what other people say about their own sleep requirements. First of all, they might be unaware of their own sleep deprivation since sleepy people are often the worst judges of their own inadequate alertness. Second, others may actually need a different amount of sleep than the average person because there are known individual differences in sleep needs just as there are individual differences in dietary requirements, activity levels, and pretty much everything else.

What Sleep Habits Will Ensure I Get the Most Out of Every Sleep Opportunity?

Once you commit to setting aside that full seven to eight-hour sleep period every night, it is important to develop good sleep habits that will maximize the benefits of each sleep opportunity. Many people know they should get more sleep, and they try their best to make it happen, but despite their efforts, something prevents a good, restful sleep experience. If you find you have chronic sleep-initiation problems in the absence of sleep disorders, circadian disruptions, or environmental factors, or if you just don't seem to sleep as well as you think you should, it is essential to implement a program of behavior change that (1) avoids practices that disturb the body's clock; (2) eliminates habits that create and maintain inappropriate associations between the bedroom and anything that is not conducive to restful sleep; (3) takes advantage of natural sleep promoters; and (4) avoids natural or chemical sleep disrupters. Each tool is listed in Table 10.1.

The details about each of these tips are discussed below. If you have sleep difficulties, adhering to these suggestions is essential (even if you do not see how these things could possibly solve your problem). Just take our word for it until you give these things a try for at least one full month.

Table 10.1 Sleep habits that lead to better sleep

- Stick to a consistent wakeup and bedtime every day of the week;
- use the bedroom only for sleep and sex;
- resolve daily dilemmas outside of the bedroom;
- establish a bedtime routine;
- establish an aerobic exercise routine and stick to it;
- create a quiet and comfortable sleep environment;
- don't be a clock watcher;
- avoid the use of mobile phones and computers close to bedtime;
- don't consume caffeine within four hours of bedtime;
- don't use alcohol as a sleep aid;
- don't take naps during the day (if you have trouble sleeping at night);
- don't smoke cigarettes immediately before bedtime;
- get out of bed and go to another room if sleep doesn't come in 30 minutes.

Stick to a Consistent Wakeup and Bedtime Every Day of the Week

Adhering to a consistent sleep/wake schedule every day of the week may be impossible for many pilots because of unpredictable flight times, shift work, and other factors. However, whenever possible, changes in the sleep schedule should be minimized. The problem with schedule variability can best be illustrated by focusing for a moment on someone who works a fairly consistent routine. Many regular day workers fall into the habit of sleeping much later on the weekends than during the week. Although this seems like a luxury that should not be missed, it often creates body-clock disruptions that turn into sleep problems. Someone who usually wakes up at 05:00 on work days, but sleeps until 09:00 on days off, is readjusting their circadian rhythm for two days every week. The later wakeup time leads to later daylight exposure and an alteration in the homeostatic sleep drive, which ultimately delays the next sleep period. All of this is fine until it is time to return to work. Now when bedtime rolls around, sleep initiation is difficult (due to the late scheduling on the previous two days) and this translates into less sleep during the night (since the work schedule dictates an early wake up). When the alarm clock sounds, the body thinks it still should be allowed to sleep an hour or two longer (since this was the case only yesterday), but it is time to get out of bed anyway, even though the circadian rhythm is still signaling sleep. As a result of the shortened sleep period and the early arise time, both the homeostatic and circadian mechanisms are "out of sync" with the work schedule, and fatigue-related mood, alertness, and performance problems result. No wonder Mondays are such a pain! Of course, everything is back to normal in a couple of days, but then the whole process starts over the next weekend if the sleep/wake schedule is disrupted again.

Use the Bedroom Only for Sleep and Sex

Using the bedroom only for things that are compatible with sleep is something that everyone should strive to accomplish. Even deployed military pilots can avoid creating non-sleep-conducive mental associations with the sleep arrangements if this is a priority. The environmental and scheduling factors may be beyond control, but playing cards, having heated discussions with your partner (in person or on the phone), or doing paperwork in bed is something everyone can avoid.

Resolve Daily Dilemmas Outside of the Bedroom

Resolving "worry issues" outside of the bedroom is tough for anyone who has the usual burdens of life and work responsibilities. However, there are some simple techniques that can help minimize the time spent lying awake in bed worrying about tomorrow. It is important to recognize whether or not you are a worrier, and while acknowledging that it may not be something that is likely to change, it nonetheless can be minimized in the bedroom.

Before going to bed, make a "worry list," and write a brief action item beside each concern. This can eliminate the process of trying to make a decision about something when the focus should be on going to sleep. Another helpful trick is to pick up the phone and call your own answering machine at work to leave yourself a message about what you intend to do about the problem next time you are in the office. Email can accomplish the same thing. None of these actions will solve the problem itself, but they can bring a sense of closure regarding a reasonable course of action that sometimes can stop the process of rethinking the same situation over and over, all night long.

Establish a Bedtime Routine

Engaging in a consistent pre-bedtime routine is beneficial for reasons that should be obvious by now. Remember, humans are creatures of habit, and once well-learned sequences of behaviors are established, one action usually automatically stimulates the next action in the chain. Because of this, it is important to adhere to a constant nightly routine whenever possible. An example would be to turn off the TV at 21:00, take a hot shower, lay out a new uniform for the next day, read a relaxing book for 30–45 minutes, set the alarm clock, and go to bed by 22:30. If this routine is followed night after night, the sequence will prepare the body and mind for the upcoming sleep, and ultimately, increase the probability that sleep will occur at the desired time. Remember, habits take a relatively short time to establish, but a long time to break. So if your bedtime routine involves behaviors that are not conducive to sleep, it will take a while to break these connections and reestablish a positive routine that leads to good sleep.

Establish an Aerobic Exercise Routine and Stick to It

The importance of aerobic exercise to restful sleep has been clearly established in the scientific literature. Studies have shown that activities such as running, cycling, and swimming during the day will subsequently make it easier to fall asleep and stay asleep during the night. Weight-lifting exercises are not as helpful, but they are better than nothing. The only caveat about exercise is that the exercise period should not be too close to bedtime because physical activity is known to have a short-term alerting effect. The rule of thumb is that no exercise usually should be performed within three to four hours of bedtime.

Create a Quiet and Comfortable Sleep Environment

Of course, creating a quiet, cool, dark, and comfortable sleep environment is important for the best possible sleep. Although complete environmental control is difficult to accomplish, it is worthwhile to at least control everything that can be controlled. Make sure the room is dark with a comfortable temperature. It is better to have the room a little cooler than normal with enough bed covers to stay comfortably warm. Noise can be masked through either a fan or some other device that will create a steady low white noise. Televisions or radios are not good sources to cover outside noises. Neither are those little electronic devices that produce wave sounds, jungle sounds, birds chirping, and so on. The exception is that tuning a radio to a position that produces constant static (that is, in-between stations) can be useful for covering noises. Intermittent bursts of sounds will interfere with sleep, whereas relatively constant sounds (a fan, an air conditioner, a waterfall, and so on) are helpful precisely because they mask intermittent noises from outside the bedroom. Ear plugs are also helpful for attenuating outside noises, although it may take up to a week to get used to sleeping with these if you are not already in the habit of using ear plugs. Make sure the mattress and pillow are firm and comfortable. Many of us try to make one mattress last for life, but these things do wear out! One should seriously evaluate whether or not the current mattress still offers the right degree of comfort and support. Most mattresses should be replaced at least every ten years, and many should be rotated on a regular basis. Pillows are also a source of discomfort if the firmness and height are not optimal. It is always worth spending a few dollars for the sake of long-term improvements in comfort and sleep.

Don't Be a Clock Watcher

The caution against being a "clock watcher" is especially important for someone who is already worried about their sleep. Checking the time throughout the night may be one of the most difficult habits to overcome, but it is very important to fight this temptation. Watching the clock sets up a maladaptive pattern of thinking that can destroy the chances of getting enough sleep. You wake up for a few seconds (like every other adult in the world), but instead of just turning over and going back to sleep, you look at the clock. Then you think "it's already midnight and I'm still not sound asleep!" Then you start to worry "will I be able to go back to sleep?" and "what if I can't fall asleep again soon?" and so on. Needless to say, this can rapidly develop into a full-scale exercise in worry and futility that really, as they say, "makes a mountain out of a mole hill." Knowing what time it is will not improve the quality of sleep, it will not make it easier to go back to sleep, and it will not increase the amount of available sleep time. So, don't do it! If necessary, put the alarm clock on a table that is out of reach, and make sure it is facing the wall where it cannot be seen. Then resolve to just go back to sleep when these brief awakenings occur. Even if it is only 15 minutes before time to get up, 15 extra minutes of sleep is better than 15 minutes of worry any day!

Avoid Technology Close to Bedtime

People need to be particularly wary of using technology in the bedroom and even within an hour of going to bed. Research suggests that electronic media exposure enhances alertness either because the bright light emitted by devices like phones, computers, or TVs suppress melatonin levels and/or because the content presented on these devices is engaging and exciting. Either way, sleep and technology don't mix!

Don't Consume Caffeine within Four Hours of Bedtime

Advice regarding the need to avoid caffeine may seem patently obvious to most readers, but it is mentioned here for several reasons. First of all, caffeine is no doubt the primary chemical fatigue countermeasure among pilots because it works, and there are no prohibitions against its use. As a result, caffeine consumption is widespread. Secondly, caffeine is known to exert a negative effect on sleep quality if taken too close to bedtime despite protests on the part of chronic caffeine users that this is not the case. The suggestion here is to avoid caffeine within four hours of bedtime, but you might be surprised to learn that a recent scientific paper indicated even morning caffeine consumption can negatively impact the nighttime sleep of some people. Thirdly, the impact of caffeine on sleep quality changes with age. Actually, the actions of the stimulant itself do not change, but since sleep architecture becomes more fragile with age, an amount of caffeine that had little effect at the age of 20 may wreak havoc on sleep quality at age 45. Fourthly, there is caffeine in a number of products that may not be noticed by the casual observer. Some brands of orange soda (and other non-cola soft-drink flavors) contain added caffeine, and many over-the-counter medications contain caffeine. The avoidance of stimulant-induced sleep problems requires careful consideration of caffeine's effects and a careful review of the labels on any foods, drinks, or medicines that could potentially contain this compound. Also, remember that tea and even chocolate contain caffeine.

Don't Use Alcohol as a Sleep Aid

The warning about alcohol's effects may come as a surprise to many because alcohol has long been considered a "sleep promoter." In fact, it is true that alcohol makes most people sleepy, and therefore increases the speed of sleep onset. However, the problem with beer, wine, and cocktails is that they disrupt the structure of sleep, particularly during the second part of the night. The negative impact of alcohol on sleep quality combined with its effects on next-day blood-sugar levels makes it a bad choice in the anti-fatigue equation. Even if the bottle-to-throttle limits are met, the ingestion of more than two drinks (four hours or less before bedtime) is ill advised.

Don't Take Naps during the Day (If You Have Trouble Sleeping at Night)

Daytime napping can be a wonderful fatigue countermeasure in situations where adequate consolidated sleep opportunities are unavailable. In fact, the positive side of napping will be discussed in detail in the section on anti-fatigue strategies for situations involving sleep restriction (Chapter 12). However, for people who are experiencing chronic sleep problems at night due to poor sleep habits, napping during the day should be avoided for one simple reason. Napping is sleep, and sleep of any length decreases the homeostatic drive for sleepiness. This means that a nap during the day will inevitably make it harder to fall asleep at night, and someone who is already experiencing nighttime sleep problems should avoid this added complication.

Don't Smoke Cigarettes Immediately before Bedtime

Smoking cigarettes right before bedtime is one more action that falls in the category of a chemical sleep disrupter. Tobacco smoke contains nicotine, and because smoke is inhaled into the lungs, its constituents are rapidly absorbed into the bloodstream. Although nicotine is a week stimulant compared to caffeine, it should be avoided by those experiencing sleep difficulties for obvious reasons. Try not to smoke within one hour of bedtime.

Get Out of Bed and Go to Another Room If Sleep Does Not Come in 30 Minutes

A final tip about good sleep habits is that it takes a while for all of these positive behaviors to overcome the years of poor mental conditioning which has seriously impaired the sleep of some people. Because of this, anyone who is attempting to correct a sleep problem associated with bad habits needs to be patient. It may take more than a couple of weeks before noticeable improvements begin to occur. Meanwhile, it is important to avoid lying in bed awake for more than 30 minutes each night waiting to fall asleep. Eventually, better sleep hygiene (habits) will produce positive effects, but becoming frustrated at the lack of immediate results will only delay progress. As the saying goes, anything worth having is worth waiting for, and quality sleep is no exception. It takes a while to build strength and endurance after embarking on a new program of physical fitness, it takes weeks to see noticeable results from a new diet, and it takes years to obtain a quality education. So, it should come as no surprise that it will take a while to overcome poor sleep habits. In the meantime, progress should not be thwarted by lying in bed for long periods becoming frustrated by a lack of instant sleep. By the way, this advice also goes for those who normally sleep soundly, but occasionally experience problems falling asleep. When sleep does not come readily, it is important to get out of bed, go into another room, and engage in some type of quiet activity (like reading or listening to relaxing music) until feelings of sleepiness return. Then, go back to the bedroom and try again. If for some reason a second or third attempt does not work, try sleeping on the couch in another room. Although this is not an optimal solution, it is certainly preferable to falling asleep in the cockpit or feeling so tired and irritable that life is miserable tomorrow. Eventually, good sleep habits will work magic, but it may take a while.

Good sleep habits are the foundation for good sleep. However, if you adhere to all of these suggestions and your efforts do not lead to better sleep within a couple of weeks, consult a sleep specialist for help.

Top Ten Points for Sleep Optimization

- The first step toward maximizing on-the-job alertness is the optimization of pre-flight sleep.
- The majority of people require seven to eight hours of restful sleep to be their best.
- Maximize the quality of off-duty sleep by implementing good sleep habits.
- When possible, stick to the same bedtime and wakeup every day (even days off).
- Keep behaviors that are not consistent with sleep out of the bedroom (this includes the use of phones, computers, and e-readers).
- Establish a consistent "getting ready for bed" routine and stick with it every night.
- Engage in aerobic exercise during the day to aid nightly sleep.
- As much as possible, make sure the sleep environment is comfortable, cool, quiet, and dark.
- Guard against the negative effects of smoking, caffeine, and alcohol on sleep quality.
- Remember, optimized off-duty sleep is the best way to maximize on-duty performance.

11 Anti-fatigue Strategies for Shift Lag and Jet Lag

The alertness, mood, and performance problems associated with irregular shift schedules and night work in particular are well known, and because of this, an obvious solution for operator fatigue would be to *avoid night work altogether*. Of course, this is a laughable suggestion in today's fast-paced, just-in-time, 24/7 world, but the idea does deserve at least some consideration. Maybe there are one or two readers out there who have enough flexibility in their operations to beg the question of whether the benefits of shift work are really worth the discomforts and risks. In situations in which around-the-clock operations provide only a slight economic or productivity advantage, the increased risk of errors, heightened levels of employee turnover, or other shift-work problems may make other alternatives more attractive.

Earlier, it was shown that when aviators participating in a research study were deprived of sleep, flight performance declined significantly (especially at night) as a function of both the sleep loss (homeostatic factor) and the body's clock (circadian factor). However, these effects are not limited to the laboratory and they are not limited to aviation-relevant tasks. Instead, it has been well established that even the simplest jobs suffer from degraded alertness at night, for the most part because of the high nighttime pressure for sleep. Figure 11.1 shows that a task as simple as reading a meter is susceptible to high error rates in the early-morning hours. As an aside, it has been found that performance also tends to suffer in the early afternoon in accordance with the sleepiness associated with the "post-lunch dip," (another circadian-rhythm phenomenon). It is no wonder that industrial accidents and automotive crashes increase during the late-night/pre-dawn hours and again in the early afternoon.

Obviously, people who have to work at night must be prepared to deal with the alertness and performance problems that are characteristic of their less-than-optimal work schedules. However, this is not the only problem that people who work non-standard hours must face. No matter how tired and sleepy people get from 03:00–6:00, when they finally get a chance to sleep, they often cannot. Everyone who has worked those "red-eye" flights knows the frustration of struggling to stay awake with

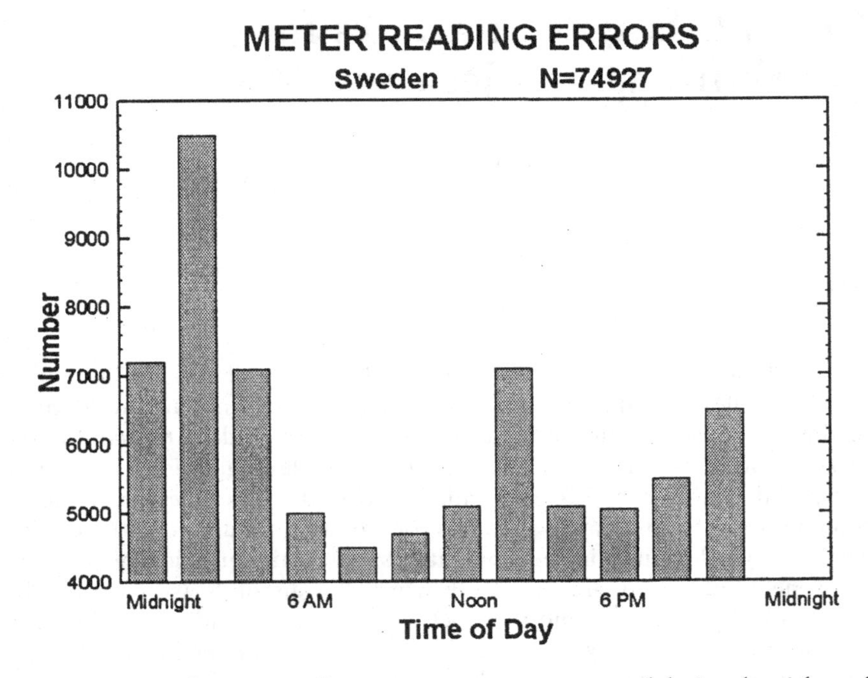

Figure 11.1 Simple meter-reading errors are more pronounced during the night and the post-lunch dip than during other times of the day

Source: Reprinted with permission from Mitler, M.M., Carskadon, M.A., Czeisler, C.A., Dement, W.C., Dinges, D.F., and Graeber, R.C. (1988), 'Catastrophes, Sleep, and Public Policy: Consensus Report,' *Sleep*, Vol. 11, pp. 100–109.

coffee, cold air on the face, conversation, squirming around in the seat, and so on, just to get home or to the hotel, fall into bed, doze off and … wake up only two hours later. What a frustration! How can this happen? Surely, everyone else does not have the same problem, or do they? Well, it turns out that the body's circadian rhythms are at work again. Except this time, instead of exacerbating sleepiness when you have to be awake, they are encouraging wakefulness when you need to be asleep. As we discussed earlier, it has been scientifically established that sleep is easier to initiate and maintain at some times of the day than at others. Going to sleep late at night generally provides a seven to eight-hour sleep period, while trying to sleep during the day provides a much shorter sleep duration and is a setup for sleep deprivation.

Of course, this turns out be more than just an irritation, especially for pilots who frequently work irregular day/night schedules and who routinely fly those transoceanic routes. Right after changing to the night shift, these folks can barely stay awake because their homeostatic and circadian drives

Table 11.1 Suggestions for dealing with night work

- When possible, reduce mental demands on night flights.
- Be sure that crews work hard to double-check everything.
- If there is scheduling flexibility, avoid long shifts since fatigue at night is already problematic (this is especially true of shifts that are long because of administrative tasks combined with flight duties).
- When possible, schedule layovers in ways that ensure enough time for seven to eight hours of sleep (taking into account the possible difficulties going to sleep as well as the need for "wind-down" time and travel time between the flight line and the place of rest).
- If using short split shifts, schedule off-duty times for periods when sleep is naturally easier to obtain.

are making them so sleepy, and then, since they cannot sleep when they finally get to bed, they are in even poorer shape the next night because they are even more sleep deprived. What is worse is that about the time the body begins to adjust to the new schedule (assuming they can actually work the same schedule for a week or more), it is time to flip back to daytime work. No wonder shift workers tend to feel tired most of the time.

Fatigue Countermeasure Strategies for Shift Workers

So, what is to be done? No amount of effort or planning will change the physiological basis for the 24-hour rhythms that affect alertness and performance, and it is clear that humans are not particularly good at being highly alert at night no matter how hard they try. Because of this, it would be best to simply eliminate job demands at less-than-optimal times and to schedule personnel to work at the times that are naturally conducive to accuracy, vigilance, and productivity. But as we said before, this will never happen. Thus, a compromise must be reached whenever possible. Table 11.1 lists some suggestions for dealing with night work, and afterwards, each tip is discussed in more detail.

Carefully Schedule Night Duty Tasks

Night duty should be kept as simple and straightforward as possible because sleepiness will impair cognitive abilities and motivation. Focus just on the task of flying the plane and avoid other distractions until a more optimal time. Remember that everyone on board is likely to be less "clear-headed" than usual, so do not assume that some unusual decision or action, or the omission of some routine procedure was actually intended. Double-check it! Try to provide shorter continuous periods at the controls if at all possible, and once on the ground, make every effort to minimize the amount of time spent taking care of "administrivia" or any other tasks that really could be put off until later. It is one thing to work long stints on daytime schedules, but prolonged work periods are very difficult for those who are already struggling to stay awake at night (or right after a night shift). Managers, commanders, and schedulers should be especially sensitive to this issue. Do not make the night workers stick around in the morning when they should be trying to get some sleep! Just because you are alert and well rested (since you just got out of bed a couple of hours earlier) does not mean that your night workers feel the same way. That staff meeting may be important, but is it important enough to compromise the safety of the next night flight? What about the safety of employees trying to drive home after work?

Attend to Circadian Factors When Scheduling Layovers

Scheduling layovers is another issue that is problematic but should be carefully considered. If you are scheduling domestic or international routes or military airlift operations, remember that the circadian rhythms of crewmembers will influence their ability to initiate and maintain restorative sleep. While delivery deadlines, maintenance requirements, and other drivers of flight schedules are important, the well-being of the man or woman in the man–machine equation needs some attention too. Circadian factors associated with time zone crossings can significantly impair the ability of crewmembers to go to sleep and stay asleep during layovers. This means that they may need a minimum of 14–16 hours off just to get seven to eight hours of sleep. Although it is not always feasible to provide this much time, it may be beneficial whenever possible.

Regional carriers, corporate/executive pilots, and military aviators involved in relatively brief sorties, usually within the same time zones, also deserve attention. Progress is being made to ensure that these folks get the needed amount of time off every day, but further work is needed to make sure that their time off coincides with a circadian phase that is likely conducive to sleep. In split-shift operations, it is great to provide some time to grab a few hours of restorative slumber, but if the sleep opportunity occurs at 17:00 in the afternoon, it will not be of much use since this is a time of day when people are normally most alert. Instead, it is best to schedule the break for the early afternoon in the "post-lunch dip," the pre-dawn hours, or the late-night hours when sleep is actually possible.

Maximize Every Sleep Opportunity

If sleep is temporarily difficult to obtain when arriving in a new environment, there are a few simple strategies that can help. If possible, an effort should be made to change the new surroundings into something more familiar by taking along a family picture, your own pillow, or something else that might be a good reminder of home. Next, it is worthwhile to spend a few minutes exploring the new environment before going to bed. Find out the location of restaurants, the fitness center, or any meeting facilities that might be needed the next day. Finally, it is a good idea to set the alarm clock for the next morning and to call the front desk for a wakeup call. Redundancy along these lines can prevent needless anxiety about oversleeping the next day. The bottom line – make the environment more familiar and eliminate as many sources of worry as possible. Doing so will maximize the potential for a restful night (or day) of sleep.

Also, take some time to optimize the sleep setting. Obviously, an uncomfortable sleep environment will not be conducive to restful sleep even when all sorts of other measures are taken to make the "home away from home" a decent place. Deployed military pilots are all too familiar with the problems of too much light, too much noise, and a sleeping arrangement that is either too hot or too cold. Unfortunately, it is difficult to avoid many of these problems in a field setting. For instance, in the 1990s during Operation Desert Storm, a number of US Army pilots were either sleeping in parking garages or tents in the desert climate of Southeast Asia. Neither setting was comfortable from the standpoint of light, noise, or heat, but there was little that could be done to correct these problems. In such situations, personnel must take advantage of individualized strategies such as the use of sleep masks to block out light and the use of foam earplugs to minimize disturbances from external noise.

Commercial aviators spending their layover sleep periods in hotel rooms have more flexibility. In a hotel, it is usually possible to control the temperature and the lighting, and a few strategies are available to deal with unwanted noise. Upon arrival it is very important to ensure that the room is as dark as possible – dark enough that you cannot see the hand in front of your face! Making sure the drapes are fully closed and the blinds are completely drawn can help to reach this goal. Remember that light is the body's primary cue to wake up and start the day, so it is especially important to avoid light exposure when attempting to sleep during daylight hours. Once the room has been made as dark as possible, measures should be undertaken to minimize noise disruptions. This means turning on a fan or air conditioner to mask out the sounds of street traffic, hallway conversations, television sets or radios in adjoining rooms, and even the sounds of running water from a neighbor's shower. Note that even small noises can have an insidious effect on sleep quality despite the fact that they do not produce a complete wakeup. Studies have shown that noise exposure can disrupt normal sleep patterns by causing a shift from a deeper to a lighter stage of sleep. If this happens repeatedly throughout the night, the restorative value of sleep will be impaired, and this will lead to feelings of fatigue throughout the next day. Thus, noise should be minimized in any way possible. Once an adequate amount of peace and quiet has been established, the temperature of the room should be adjusted so that the environment is cool enough to be very comfortable once underneath the bed covers. Obviously, a room that is either too cold or too hot should be avoided since the physical discomfort associated with less than optimal room temperature will adversely affect the quantity and quality of sleep. Studies show that temperatures between 60 and 67 degrees Fahrenheit are optimal for sleeping; if temperatures are above 75 degrees or below 54 degrees, sleep is disrupted. A cool, dark, and quiet bedroom is essential to ensure maximum restoration before the next duty period.

Circadian Adaptation Strategies for Sufferers of Shift Lag and Jet Lag

For people who work irregular schedules or those who routinely cross multiple time zones, disruptions to the body's clock are a significant problem. It has been estimated that it takes (at best) one full 24-hour period to adjust to each one-hour change in work/rest schedule. Thus, full adaptation to a nine-hour work/rest schedule change or a nine-hour time zone shift will take approximately nine days. Of course, the direction of time change can affect the speed of adaptation, with forward clock rotations (westward travel) showing faster adaptions than counterclockwise (or eastward travel). In both situations, the circadian rhythm is disturbed, and this results in insufficient sleep, increased fatigue, and sleepiness on the job. People end up working at times when their bodies are programmed for sleep, and sleeping (or trying to sleep) at times when their bodies are programmed to be awake, and such schedules are contrary to our basic physiology.

Remember what we have already said about the fact that humans were sleeping at night and working (or active) during the day, within the same time zone, for thousands of years before the 1883 invention of the electric light bulb effectively "turned night into day," and the 1950s saw an increased reliance on night-fighting capability due to the advent of night-vision technology. Couple these developments with the fact that modern aviation technology has made it feasible to rapidly traverse multiple time zones, and it becomes easy to see that present-day work demands have outstripped our basic biological ability to keep up.

In 1913 it took *nine days* to cross the six time zones between the US and Europe on the fastest transportation available (a steam liner), but now this same trip takes only *seven hours* in a commercial Boeing 747 or a military C-141B or C17. From 1913 to 1964, technology increased the speed with which we could move from one time zone to another from a low of one zone per day to today's rate of one zone per hour. Meanwhile, the rate at which we humans can adapt has not changed! Our rate of adjustment is a minimum of one time zone per day. So in 1913, we were keeping up with transportation technology just fine, but in 2015, we are falling behind. No wonder we need help to minimize the sleep disruptions and performance and alertness decrements associated with misalignment of the body's internal rhythms, and even from the requirement for personnel to awaken at inopportune times.

Determine Whether a Circadian Readjustment is Really Necessary

Fortunately, there are a number of strategies that facilitate changes in the body's circadian rhythms. However, before embarking on a program to readjust the internal clock, take a moment to consider the appropriateness of the effort. A decision must be made about whether it is worth it to fight the circadian rhythm at all. If only one to three days will be spent in the new work shift or time zone, the best strategy is to try to remain on the original schedule. This may mean eating and sleeping at times that are out of phase with the new time shift, but unless there is some compelling reason to do otherwise, this is preferable to the discomfort, fatigue, and sleep deprivation that will result from attempting to readjust the body's rhythms for such a short time.

If it looks like more than three days will be spent in the new shift or time zone, a planned adaptation routine should be undertaken. This plan should include a number of carefully-planned, systematic techniques.

Properly Rotate Schedules

Schedulers need to set up shift rotations that offer personnel the best possibility for adjusting to new work/sleep times. Whenever possible, forward or *clockwise* rotations should be used. This means starting with the normal dayshift, then moving to an evening shift, and finally to a night shift rather than the other way around. This is much easier for the body to handle because it takes advantage of our natural ability to delay sleep until later than usual as opposed to fighting our natural inability to fall asleep earlier than usual. This "forward rotation" strategy takes advantage of the same physiological propensities that make it easier to readjust to a westward time zone change than to an eastward transition.

Use Medications When Necessary

Sometimes the only effective way to induce sleep at a time that is contrary to the body's clock is through the use of prescription sleep medications. There are certain situations in which these medications should definitely be considered if they are authorized by appropriate regulations and by the leadership. Modern prescription sleep aids are effective, short-acting, and largely devoid of the problematic side effects that were at one time associated with sleeping pills (and in both civil and military operations, the use of these medications is sometimes permitted as long as the proper grounding times are observed).

Recommendations concerning which sleep aid is most appropriate must take into account the nature of the situation in which sleep is desirable. There are two main categories of hypnotics (sleep aids) that are possibilities – short-acting and intermediate-acting. Short-acting hypnotics are appropriate for inducing naps because most of their effects will dissipate after only two to four hours. Thus, these hypnotics will permit the advantages of a short sleep period without creating significant problems with "hangovers." Common choices are zolpidem (Ambien®), triazolam (Halcion®), and zaleplon (Sonata®). A study we conducted in 1998 indicated that when one of these drugs was given to aviators who were trying to sleep at an inopportune time, it produced better and longer naps than an inactive placebo, and as a result, subsequent alertness, mood, and performance were facilitated (see Figure 11.2).

Short-acting sleep medications also are great for people who are trying to fall asleep earlier than usual at night in preparation for an early report time. The reason for this is that short-acting compounds will promote rapid sleep onset, and they can maintain sleep until after the body's normal bedtime has passed (at which time natural mechanisms will take over). Then, their effects will subside long before it is time to wake up for work.

Intermediate-acting hypnotics are more appropriate for inducing and maintaining the daytime sleep needed by shift workers, or the sleep of people who have recently crossed multiple time zones. The most popular one of these medications is temazepam (Restoril®). Temazepam will promote sleep within about 30 minutes; it lasts for a long time, which maintains sleep even during periods that are contrary to the body's circadian cycle. These characteristics make it a good choice for maintaining sleep throughout an eight-hour period. One investigation that we completed in 2002 showed that when temazepam was used to promote the daytime sleep of pilots who were shifting to a night schedule, the pilots who received the drug slept normally whereas those who received the placebo got less sleep and experienced more frequent sleep disruptions. Due to the temazepam-induced improvements in daytime sleep, simulator flight performance and alertness on the second night of night work were improved compared to the group who received placebo (see Figure 11.3).

Temazepam is superior to zaleplon, zolpidem, or triazolam for daytime sleepers or for those who are suffering from jet lag simply because people in these situations suffer from difficulties *staying asleep* rather than problems *going to sleep*. The longer-lasting effects of temazepam are extremely beneficial in such circumstances, because they tend to prevent premature awakenings. Remember that all these medications are prescriptions and must be administered by a physician. In aviation environments, aircrew must stay informed of what medications are authorized and consult with their flight surgeon or Aviation Medical Examiner (AME) on the appropriate use of these medications.

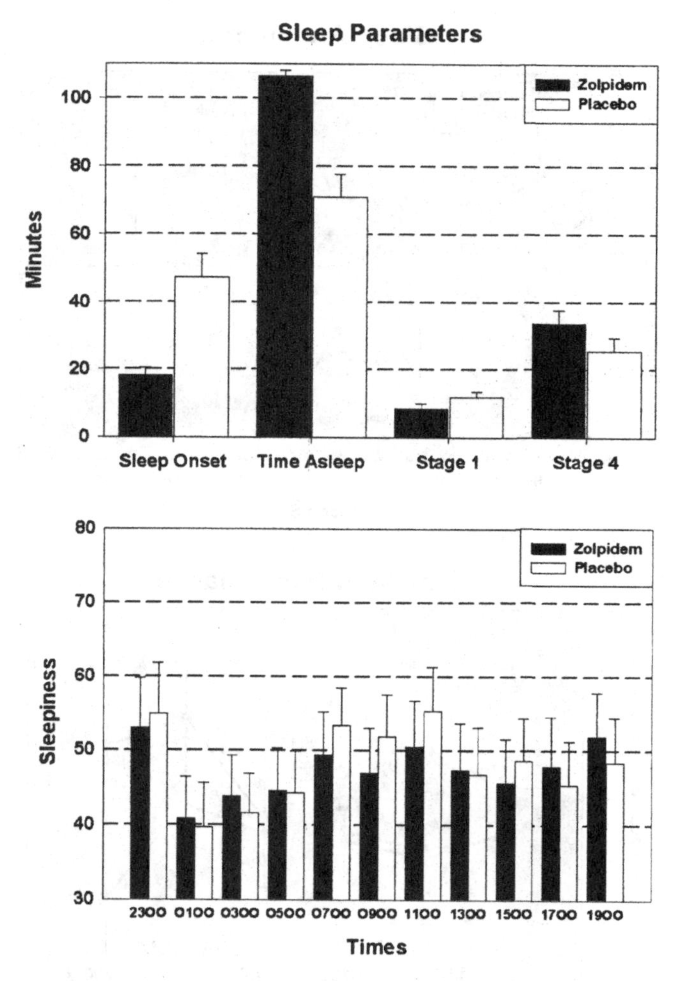

Figure 11.2 A zolpidem nap is better than a "natural nap" when trying to sleep at an inopportune time

Source: Caldwell, J.A., Jones, R.W., Caldwell, J.L., Colon, J.A., Pegues, A., Iverson, L., Roberts K.A., Ramspot, S., Sprenger, W.D., and Gardner, S.J. (1997), 'The Efficacy of Hypnotic-Induced Prophylactic Naps for the Maintenance of Alertness and Performance in Sustained Operations,' *USAARL Report No. 97–10*, US Army Aeromedical Research Laboratory, Fort Rucker, AL. (Public Domain).

Employ Behavioral Interventions Combined with a Few Other Tips

When fighting shift lag and jet lag, it is essential to quickly adjust meal, activity, and sleep times to the new schedule. Adjust the timing of sunlight exposure in accordance with the guidance provided below. To overcome

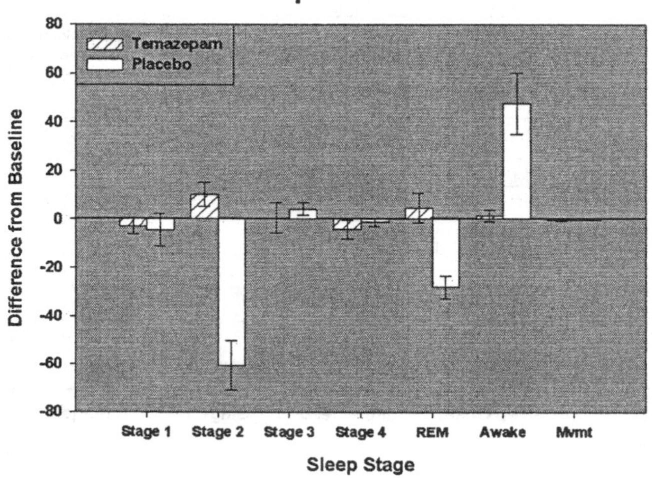

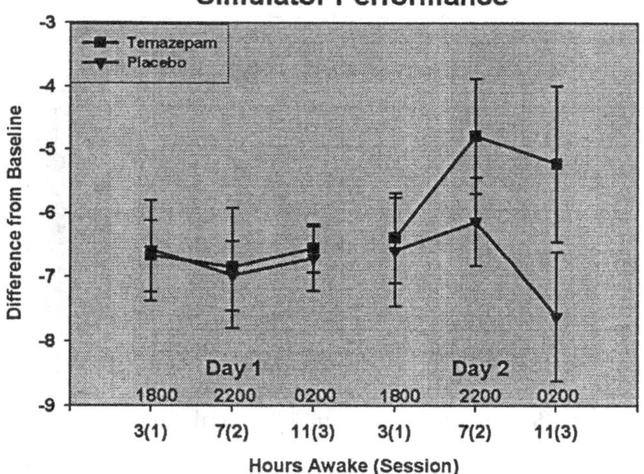

Figure 11.3 Temazepam can improve night-shift flight performance by creating better daytime sleep

Source: Caldwell, J.L., Hall, K.K., Prazinko, B.F., Norman, D.N., Rowe, T., Erickson, B.S., Estrada, A., and Caldwell, J.A. (2001), 'The Efficacy of Temazepam for Improving Daytime Sleep and Nighttime Performance in Army Aviators,' *USAARL Report No. 2002–05*, US Army Aeromedical Research Laboratory, Fort Rucker, AL. (Public Domain).

sleep problems, make sure the sleep environment is optimal in accordance with the suggestions presented earlier. Avoid heavy meals before bedtime since these may create sleep-disrupting gastrointestinal disturbances. Try a hot bath and relaxation exercises right before bedtime. Implement all of the guidelines for good sleep habits (see the previous chapter). And finally, be prepared to use caffeine upon awakening to help overcome the grogginess that will result from a shortened sleep period and, from the standpoint of the body's clock, an early-morning awakening. In time, the body's clock will somewhat adjust to the new schedule, sleep will get better, and daytime fatigue will gradually diminish.

Try Bright Light

The use of bright lights may help to minimize some of the problems associated with schedule changes by promoting a more rapid resynchronization of the body's rhythms to new routines. It is an established fact that light exposure exerts a substantial influence on the timing of the circadian system, and research has shown that bright light can produce large phase shifts when the light exposure is timed in conjunction with a new sleep/wake cycle. To minimize "shift lag" when working irregular schedules, particular attention should be given to maximizing light exposure at certain times, while minimizing light exposure at other times. This may require special (but relatively inexpensive) tools. For instance, night workers will need portable bright-light devices to provide sufficient light exposure at times when sunlight is not available, and they will need dark wrap-around sunglasses or blue-light blocking glasses to block inappropriate sunlight exposure during the daytime. If possible, it is desirable for people who are changing from day shift to night shift to be exposed to artificial bright light for at least two hours in the evening (ideally, prior to work, beginning at approximately 20:00), but to avoid sunlight exposure in the morning on the way home from work (using dark or blue-light blocking glasses). This schedule tends to "push" the sleep period later, because it is like getting a sunrise signal at night (right before work) while avoiding the typical bright-light exposure in the morning (right before sleep). In circadian-rhythm terminology, this schedule is designed to produce a *phase delay* because the timing of the next sleep period is being forced later than normal. Night workers who are transitioning to the day shift should attempt to promote a *phase advance* (pulling the sleep period earlier than normal). These people should be exposed to bright light shortly after about 09:00, and they should of course avoid as much light as possible in the evening.

Early research indicated that the amount of light sufficient to produce beneficial effects needed to basically mimic sunlight – very bright and full spectrum. As the research progressed, evidence grew that even bright room lighting (150–300 lux) was beneficial (although this amount is not as helpful as the 1,000–10,000 lux intensities that can be generated by some artificial

Table 11.2 Bright light exposure can improve nighttime alertness and performance

- Sunny day – 100,000 lux
- Cloudy day – 10,000 lux
- Bright artificial light – 2,000 lux
- Ordinary indoor light – 200 lux

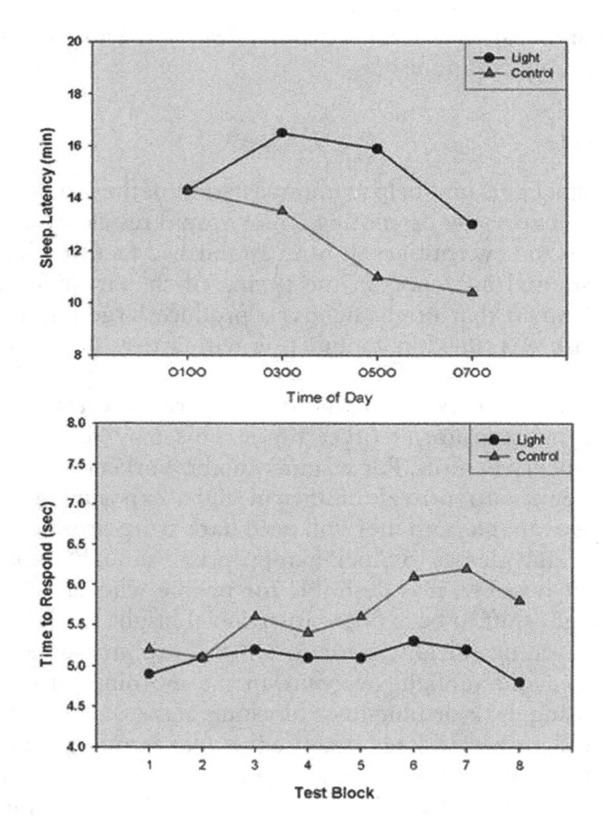

Figure 11.4 Bright light exposure can improve nighttime alertness and performance
Source: Reprinted with permission from Campbell, S.S. (1995), 'Effects of Timed Bright-Light Exposure on Shift-Work Adaptation in Middle-Aged Subjects,' *Sleep*, Vol. 18, pp. 408–416.

sources). Table 11.2 shows the typical lux levels associated with different types of lighting.

Figure 11.4 shows the benefit of bright-light exposure on sleepiness and performance. Compared to sleepiness levels and performance without bright light, the graphs indicate that people will be more alert (have a longer sleep latency) and perform better (have a shorter response time) during the night shift when a bright-light intervention is employed.

In addition to the benefits of bright light, scientists have found that the color of the light is very important as well – monochromatic light in the blue wavelength suppresses melatonin levels and exerts more influence on the circadian system than longer wavelengths of light or broad-spectrum white light in general. One study found that driving performance was better when the drivers were exposed to blue light on the dash while driving in the early-morning hours than without the light. Other research has shown better alertness and faster reaction times with blue light exposure compared to low levels of light in the evening.

Figure 11.5 shows the effects of blue light implemented in the workplace. Note the improvement in concentration, eye comfort, mood, and post-work sleep quality.

Properly timed light exposure is also important for adjusting circadian rhythms after traveling to new time zones. Generally speaking, jet lag becomes a problem after crossing three or more time zones, and the effects of jet lag are very similar to those of shift lag. However, jet lag can be easier to overcome than shift lag because the environmental cues at the

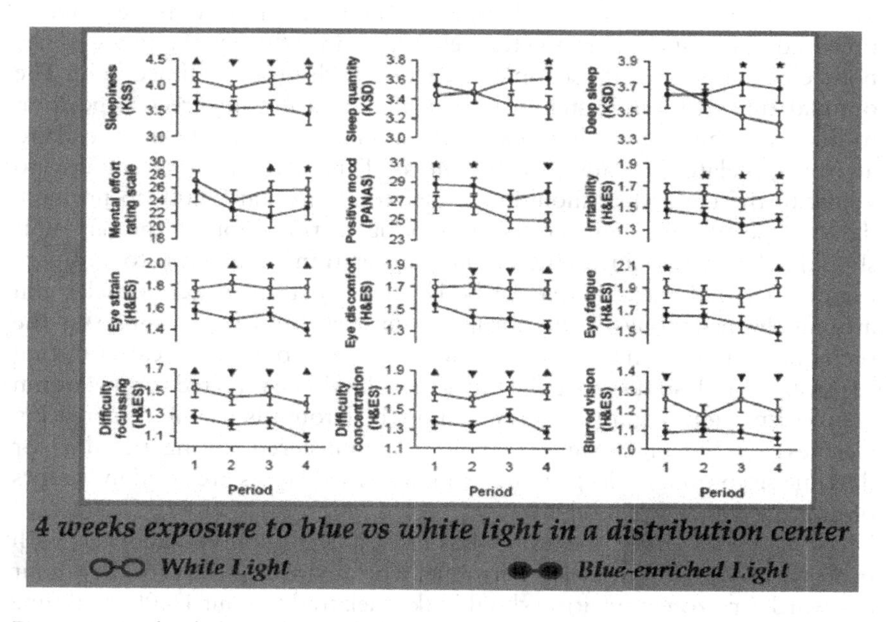

Figure 11.5 Blue light in the workplace can improve performance

Source: Reprinted with permission from Viola, A.U., James, L.M., Schlangen, L.J.M., and Dijk, D-J. (2008). 'Blue-enriched White Light in the Workplace Improves Self-reported Alertness, Performance and Sleep Quality.' *Scandinavian Journal of Work and Environmental Health* Vol. 34, pp. 297–306.

destination are more conducive to rapid readjustment. In other words, unlike shift workers who are often working against normal environmental factors, people who travel to a new time zone will naturally be able to experience sunlight during the day, and they will not have trouble avoiding too much light exposure at night. In addition, those who travel to new time zones will benefit from social/work interactions, meals, and other activities at times that coincide with their new "daytimes" while everything will be quiet and sedate at the times that coincide with their new sleep period. This is of course not a luxury that night workers have since they are trying to stay awake when the rest of society is sound asleep and trying to sleep when everyone else is up and about.

Consider Melatonin

Melatonin is a potential aid for those who are attempting to readjust their body's clock because of shift work or travel. In addition to the influence of light, the hormone melatonin, secreted by the pineal gland, has been found to be important for establishing and maintaining the timing of the circadian system. Although many questions remain about the long-term safety and efficacy of using melatonin supplements and little definitive information is available concerning optimum dosages, research has shown that the properly timed administration of synthetic melatonin can influence the phase of the body clock, especially if used in conjunction with properly timed light. The optimal times to administer melatonin are basically in direct opposition to those at which bright-light exposure should occur (bright light leads to alertness, melatonin leads to drowsiness). Thus, day crews attempting to adjust to the night shift should take melatonin around 09:00 (to promote daytime sleep), and night crews attempting to transition to the day shift should take melatonin about 20:00 (during transition days, to promote evening sleep). Remember that the body normally starts to secrete melatonin around the beginning of the normal sleep cycle, and light suppresses the secretion of natural melatonin. The levels of melatonin are greatest during the sleep period, so these are the times that should be targeted for melatonin supplementation. Figure 11.6 shows how melatonin is useful at increasing alertness on the night shift when it is administered during the day for daytime sleep (longer sleep latency and decreased lapses are good indicators of improved alertness levels).

For help with jet lag, melatonin has proven useful *if* used in concert with properly timed light exposure. Personnel who are making a six to seven-hour eastward time zone transition should take melatonin about 16:00 local time on the day of departure. Upon arrival at the destination, melatonin should be taken in the evening. The timing of melatonin is dependent upon the number of time zones crossed added to 17:00. So, if one is traveling from the eastern time zone to Rome, six time zones are crossed. When it is 07:00 EST, it is 13:00 in Rome. Based on the rule of thumb, melatonin should be

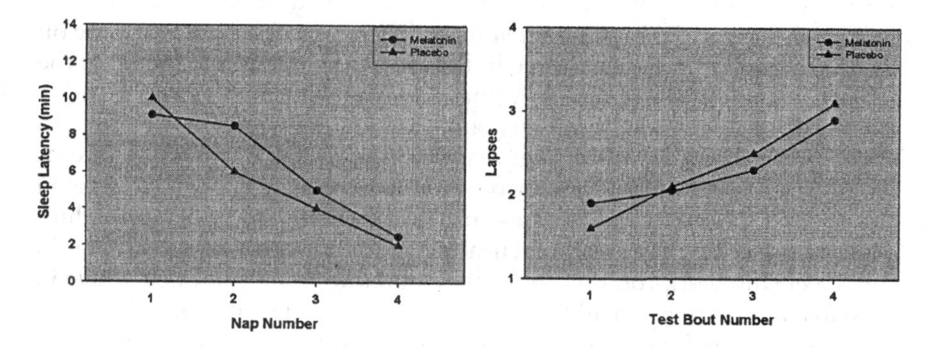

Figure 11.6 Melatonin administration can improve nighttime alertness and performance when administered for daytime sleep

Source: Reprinted with permission from Sharkey, K.M., Fogg, L.F., and Eastman, C.I. (2001), 'Effects of Melatonin Administration on Daytime Sleep after Simulated Night Shift Work,' *Journal of Sleep Research*, Vol. 10, pp. 181–192.

taken at 23:00 in Rome (1700 + six time zones is 23:00) the first night. For the next three or four nights, melatonin should be taken one to two hours earlier until adjustment to the local time occurs. If possible, a sleep aid is recommended to help promote sleep while this adjustment occurs. For westward transitions, melatonin may be taken at the local bedtime or later for up to four days after arrival. If early-morning awakenings occur, another dose of melatonin may be useful in prolonging sleep until the desired rise time, although melatonin does not have the sleep-facilitating properties of prescription sleep medications.

Aviators are cautioned that they should avoid taking melatonin prior to an overseas flight or certain other critical duty periods because melatonin may actually produce sleepiness, and a sleepy pilot at the controls is what we are trying to avoid. However, if you find that you will be staying at your destination for several days of rest, then melatonin administration may work for you. Military pilots should check with their command's policy to make sure melatonin is approved for use.

Of course, this is an extremely overly simplified discussion of very complex issues, and personnel interested in using bright lights or melatonin should be aware that there are vast individual differences in biological rhythms and other factors that may either increase or decrease the effectiveness of such approaches. The intervention times suggested above are based on assumptions about the *average* timing of peaks and troughs in the body's temperature and hormonal rhythms, and these are not consistent for everyone. The only way to ensure the proper use of light and melatonin is to evaluate each individual's circadian rhythms prior to employing such interventions, and if you are interested in using such a systematic approach, you should consult a paper by Dr Alfred Lewy and colleagues (published in 1998 in the journal *Chronobiology International*). It is crucial to minimize conflicting cues (such as artificial bright light at night followed by sunlight exposure on the way home from the flight line in the morning) in order to facilitate the efficacy of these strategies, and this may be impossible in a variety of flight operations. Furthermore, both civil and military pilots may not be able to utilize bright-light exposure immediately prior to night flights because of the adverse impact of this strategy on dark adaptation (although working in an area that is brightly lit 30–45 minutes pre-flight would be helpful). Thus, while both techniques are theoretically suitable for countering fatigue in personnel who work rotating shifts, their application may not be feasible in most aviation settings. In addition, with regard to jet lag, it should be reiterated that it is unwise to attempt shifting the circadian rhythm to a new time zone when the duration of the trip is only one to two days.

Computerized Scheduling Tools Can Help Devise New Work/Rest Schedules

Before concluding this chapter, it is worth noting that technology offers one other technique for attending to both homeostatic and circadian factors when developing new work/rest schedules. The development of sound work/rest schedules is obviously important to ensure maximum alertness on the job and optimal restorative sleep during off-duty periods. However, it can be difficult to decide which of several scheduling variations is best given the operational constraints present on any particular day. What is the best time to plan high-workload cycles? Will a one-hour nap at 02:00 be better than a two-hour nap at 05:00? How much of a dip in performance is likely to occur at 04:00 in the morning after 19 straight hours of work? Luckily, the answers to these and other scheduling questions can be determined through the use of computerized scheduling tools. At present, a variety of these tools are being developed for operational use, and some are already available for full implementation.

Models Underlying Predictions and the Effects of Individual Differences

Each tool has unique features, but most if not all use some type of two or three-process model to predict alertness. As noted earlier, the two main components are (1) the *time awake since the last sleep episode*; and (2) the *time of day*. The third major component, which is particularly important when planning strategic naps, is *sleep inertia* (or post-sleep grogginess). In addition to these three processes, various scheduling tools factor in other important influences. Some include the effects of alertness-enhancing substances such as caffeine. Others account for the effects of consecutive duty periods, the amount of time on task, and factors that impact the quality of restorative off-duty sleep. Still others take into account the effects of light exposure on both immediate levels of alertness as well as on the overall circadian system. Each of these tools offers general predictions of the levels of alertness or performance effectiveness that may be expected to result from particular types of work/rest schedules. Often times, the models are coupled with software routines that predict the amount of sleep that would be expected on different schedules, or they offer ways to interface the predictive software with sleep measured via wrist actigraphy. While the predictions are not at present capable of taking into account individual differences in sleep needs, stress tolerance, fatigue resistance, or circadian-adaptation speed, they can provide a general idea of the acceptability of candidate scheduling options. After all, so-called "group predictions" are more relevant to most business and military scenarios than "individualized predictions" since airlines and military commands generally do not have the luxury of being able to implement hundreds of individually optimized flight schedules in the operational context.

Choosing the Right Scheduling Tool

Which of the many scheduling tools is best? This question is impossible to answer with any degree of certainty right now since further field validation and fine-tuning is needed. However, these issues should not discourage the selection of a tool that looks appropriate, nor efforts to try it out in the operational context. Just remember, *the advice offered by any particular computer algorithm is intended to be a general guide* rather than something that must be followed to the letter. For our present purposes, no attempt will be made to recommend one model over another. However, an example of the sort of scheduling assistance that is now available will be drawn from the Fatigue Avoidance Scheduling Tool™ (FAST) that is currently being used in some aviation settings.

One Possibility is the Fatigue Avoidance Scheduling Tool™

The FAST™ makes predictions about the levels of performance effectiveness that can be expected with specific work/rest schedules. The predictions are based on an underlying model called SAFTE™, developed by Dr Steven R. Hursh at Science Applications International Corporation, and is diagramed below in Figure 11.7. Note how the influences of both homeostatic (time since sleep) and circadian (time of day) factors are included.

After the user has input the parameters of a specific work/rest schedule into the computer, the FAST™ generates output similar to what is shown below in Figure 11.8. The panel on the left indicates what is predicted to happen to performance during a period without sleep. The panel on the right suggests how a brief sleep period in the early-morning hours might improve performance following the nap.

In this example, only a two-day sleep/wake schedule is presented, and in this illustration there was no need for calculation of estimated sleep times.

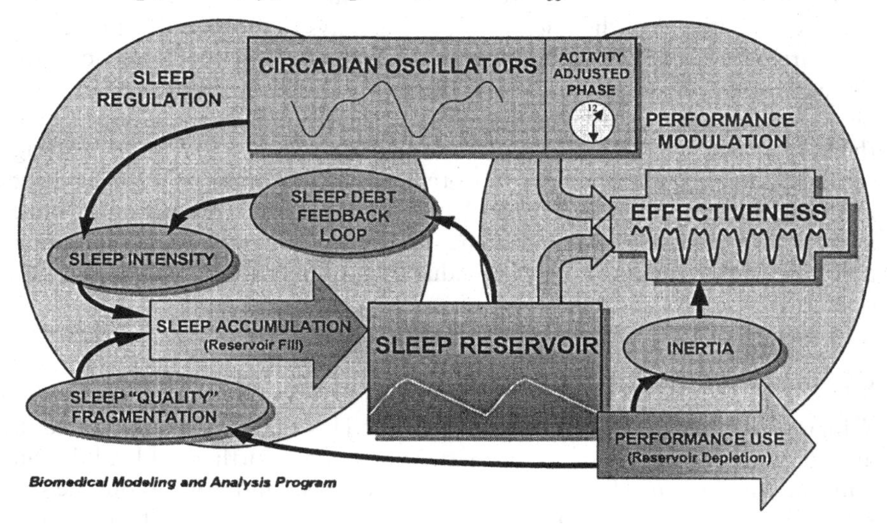

Figure 11.7 The Sleep, Activity, Fatigue and Task Effectiveness (SAFTE™) model used to predict performance effectiveness

Source: Roma, P.G., Hursh, S.R., Mead, A.M., and Nesthus, T.E. (2012), 'Flight Attendant Work/Rest Patterns, Alertness, and Performance Assessment: Field Validation of Biomathematical Fatigue Modeling,' *Report Number DOT/FAA/AM-12/12*. Federal Aviation Administration Office of Aerospace Medicine, Washington DC. (Public Domain).

But the FAST™ is capable of displaying much more complex, multi-day schedules containing a variety of consolidated sleep opportunities, a number of daytime or nighttime work schedules, and naps of various durations. In addition, there is an auto-sleep algorithm that can estimate the amount of sleep that would be expected on a day-to-day basis as a function of the timing and duration of sleep opportunities, and there is an adjustable "sleep quality" parameter, which helps to deal with the fact that disrupted sleep provides less restoration than restful sleep. Although the FAST™, or any other similar tool, cannot determine exactly how any single individual will respond to a particular schedule, these computer programs should be considered as one type of fatigue countermeasure since they can help determine which of two or more proposed work/rest routines will best facilitate the alertness levels for most personnel. Computerized scheduling tools are particularly helpful in deciding where to place the sleep opportunities available in different circumstances, and they are great for revealing the points of greatest fatigue risk in any real or planned schedule.

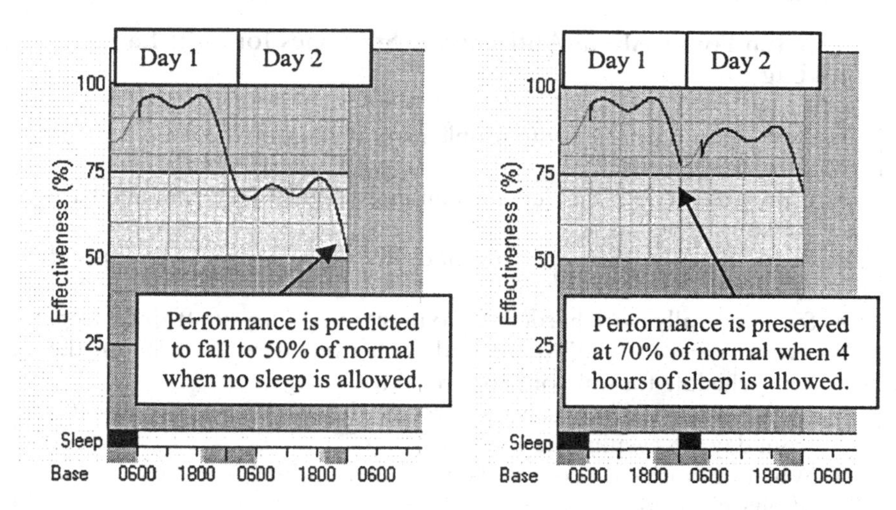

Figure 11.8 A scheduling tool (FAST™) to aid in planning work/rest hours when sleep is difficult to obtain

Also, it is now possible to augment the accuracy of the fatigue predictions generated from models such as SAFTE™ via the use of sleep data obtained from wrist-worn activity monitors. Validated wrist activity monitors overcome the disadvantage of estimating rather than actually measuring operator sleep and this, of course, improves the accuracy of model-based fatigue-risk calculations. Although actigraphy is not fail-safe because it cannot accurately detect relaxed (movement-free) wakefulness or microsleeps (that is, lapses into sleep that last for 30 seconds or less), it is far better at tracking bedtimes, wakeup times, and sleep times than are subjective sleep logs or software-based sleep estimation routines. Actigraphically measured sleep histories can provide a solid indication of risk levels for operational fatigue attributable to sleep loss and disrupted sleep/wake cycles.

In summary, there are a number of techniques that can be helpful for promoting sleep in situations where sleep is feasible, and there are several strategies that can improve circadian adaptation in operational settings. But what should be done when severe sleep restriction or total sleep deprivation is threatening performance and safety? In the next chapter, this problem will be discussed along with countermeasures designed to keep people awake for long periods of time and/or during naturally occurring troughs in operational readiness that are exacerbated by a high level of sleep debt.

Top Ten Points About Anti-Fatigue Strategies for Shift Lag and Jet Lag

- Shift lag and jet lag invite problems that can only be overcome by a focused approach.
- If possible, reduce the demands on night flights since personnel are already struggling.
- Learn the best times for the body to sleep during layovers and on-board rest breaks.
- Systematically work harder to maximize the quality of off-duty sleep.
- Avoid trying to readjust the body's rhythms if you will be on the new schedule for only one to three days.
- If changing to a new schedule, rotate the work period forward – from days to evenings to nights.
- When authorized, take short-acting medications to promote "out of phase" sleep.
- Quickly adjust meal and activity schedules to the new work/rest schedule.
- Use bright sunlight or artificial light at the right times to help resynchronize the body's rhythms.
- Consider using a computerized scheduling tool in combination with wrist actigraphy to identify the fatigue risk associated with any new or existing work/rest schedule.

12 Anti-fatigue Strategies for Situations Involving Sleep Restriction

In some settings, the opportunity for sleep is non-existent or very limited, but the job still must be completed. These situations are particularly common in military operations and emergency situations where lives may be lost if the mission is not performed. In such situations, the risk/benefit equation changes from what is normally found in day-to-day flight operations. Most of the time, flying while severely fatigued is simply unacceptable because choosing to perform under impaired conditions can only lead to problems. However, when the choice to *decline* a flight actually poses a safety problem because it places several human lives at risk, there may be considerations other than the possibility that the aircraft and crew could be in jeopardy. For instance, in firefighting operations, a pilot's choice to forgo a mission could indirectly permit the further spreading of an inferno that could endanger an entire community. In such situations, there may be little choice but to press on in hopes that everything will work out okay in the end, despite dangerously high levels of sleepiness/fatigue in the aircrew involved. This chapter will discuss strategies to improve alertness even in contexts where adequate sleep opportunities are minimal.

Strategic Napping

Well-planned naps can serve a maintenance or a recuperative function to attenuate the effects of fatigue on performance until normal sleep is once again possible. This is the best countermeasure for fatigue besides seven to eight hours of consolidated restful sleep. Napping is so effective simply because it *is* sleep, and since only sleep can address one of the two primary factors underlying sleepiness/fatigue (the homeostatic sleep drive), naps should be considered a first-line approach to preserving alertness on the job whenever seven to eight hours of consolidated sleep is not possible. Although napping exerts little or no effect on the circadian component of sleepiness/fatigue, it can mitigate the general impact of fatigue during circadian "low points" by reducing overall sleep pressure.

Every commercial pilot understands how difficult it is to obtain adequate continuous sleep in long-haul operations and during short layovers, and every military pilot understands how sleep deprivation is a fact of life in continuous and sustained military operations. Pilots employed by regional carriers and those flying corporate/executive jets or military VIP missions likewise can identify with the sleep difficulties resulting from unpredictable weather, traffic delays, maintenance problems, and scheduling difficulties. Those who work in these sectors are all too familiar with the fatigue that hits after disruptions in pre-mission sleep are followed by 18 or more continuous hours of wakefulness. That sinking feeling that accompanies the daily circadian trough in situations where drowsiness is already problematic can present a real challenge to even the most seasoned crews. In these situations, naps can offer significant relief because this is the only strategy that can help to replenish the daily requirement for sleep.

Given the right circumstances, napping can be easier to implement than precisely regimented crew schedules or rigid duty limitations, and napping has been shown to maintain or improve performance across a wide variety of settings. Canadian researchers Drs Angus, Pigeau, and Heslegrave, for instance, have found that even after 40 hours of sleep deprivation, a two-hour nap prior to an additional night of sleep loss can maintain performance at 70 percent of well-rested levels. More to the point for aviators, Dr Rosekind and his colleagues at NASA proved that 30-minute cockpit naps were able to prevent many of the attention lapses and involuntary episodes of sleep intrusion (micro-events) encountered by crews engaged in long-haul flight operations. We have found that even naps taken *prior* to a night of continuous work (prophylactic naps) can minimize alertness decrements associated with sleep deprivation. Figure 12.1 below shows that naps taken before a long, overnight work phase can attenuate or arrest performance declines.

Although two to four-hour naps are considered optimal for arresting performance declines associated with continuous work without sleep, even short naps (20–30 minutes in length) have been found to enhance the productivity and safety of sleep-deprived personnel. The bottom line is that scheduled napping of almost any duration offers a significant performance advantage over untreated sleep deprivation, and because of this, napping should be used whenever possible. However, for napping to be properly implemented in operational contexts, several factors must be considered.

Create a Suitable Nap Environment

First, it is essential to create an environment that is conducive to restful, restorative sleep. This may seem like a "no brainer" but it is amazing how many times the importance of environmental factors is overlooked in operational settings. It is essential to provide a napping area that is dark and comfortable, and one in which noise and distractions are minimal. In situations where lighting and noise are beyond control, sleep masks,

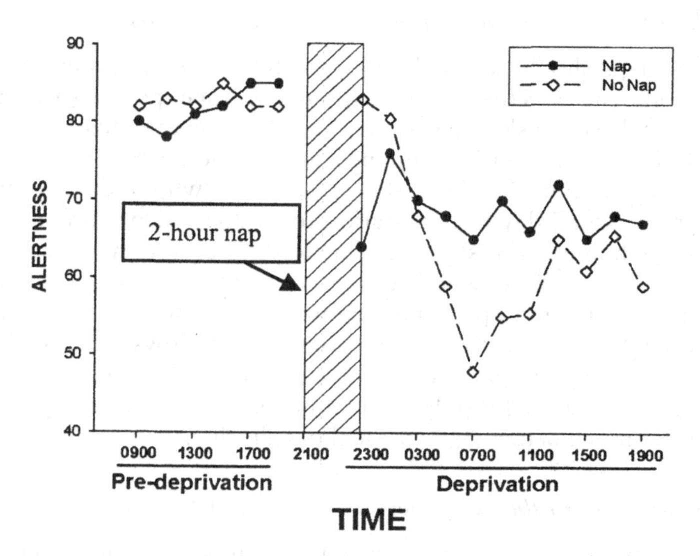

Figure 12.1 Naps can temporarily make up for lost sleep
Source: Caldwell, J.A., Jones, R.W., Caldwell, J.L., Colon, J.A., Pegues, A., Iverson, L., Roberts, K.A., Ramspott, S., Sprenger, W.D., and Gardner, S.J. (1997), *The Efficacy of Hypnotic-Induced Prophylactic Naps for the Maintenance of Alertness and Performance in Sustained Operations*, USAARL Report No. 97–10 (Public Domain).

foam ear plugs, or some type of constant "masking" sound can be used to overcome environmental sources of sleep disruption. Although slipstream noise can be fairly intense in flight, and the sound of portable generators can be heard through the walls of tents or buildings that house military pilots on the ground, these can actually be helpful since they are relatively constant and both are loud enough to hide a variety of intermittent sounds that can disrupt sleep quality.

Time Naps Appropriately for the Circadian Phase

Second, the timing of naps should be established in such a way as to promote the easy natural initiation of sleep. Since circadian rhythms affect the ability to go to sleep at different times of the day and night, whenever possible it is best to place nighttime naps between 01:00 and 6:00 (according to the body clock – and not necessarily local time) and daytime naps between 14:00 and 16:00. As discussed earlier, both of these times are associated with natural dips in alertness that will make sleep easier to initiate and maintain. Obviously, it is counterproductive to offer napping opportunities during periods that are normally associated with high levels of physiological wakefulness such as between 10:00 and 12:00 during the day or between 17:00 and 20:00 in the evening.

Time Naps Close to the Beginning of Long Duty Periods

Third, the naps should be as close as possible to the beginning of a long duty period. The effectiveness of naps lies in their power to reduce the homeostatic drive for sleep (remember that the primary factor determining the level of the homeostatic component is the amount of time awake since the last sleep period). So, for instance, someone who wakes up at 06:00 and stays awake all day in preparation for a 23:00 departure already has a significant homeostatic sleep drive before he or she even performs the pre-flight. Already, there have been 17 hours awake before wheels up, so by time to land, the total period of continuous wakefulness can easily reach 26–30 hours! Compare this to someone who follows exactly the same schedule with the exception of grabbing a two-hour nap at 14:00. This person has been awake for only seven hours prior to takeoff, and as a result, he will be in much better shape throughout the flight.

Place Naps Early in the Sleep Deprivation Period

Fourth, in situations where the work demands make sleep loss unavoidable, naps should be placed relatively early in the sleep deprivation period since it is always easier to *prevent* fatigue-induced decrements than it is to *restore* performance that has already deteriorated. Even powerful prescription stimulants must be given at much higher doses to bring severely fatigued people back to well-rested levels, whereas much smaller prophylactically administered doses can prevent many problems from ever appearing in the first place. Of course, it may be necessary to subsequently arrange for additional naps to prevent performance decrements later on.

Make Naps as Long as Possible

Fifth, the naps should be as long as possible, while keeping in mind that at least 20 minutes should be allowed for sleep inertia (post-sleep grogginess) to dissipate before returning to flight-related duties. Earlier, the dose-response relationship between the amount of sleep and the degree of alertness was discussed. People who are provided with the opportunity to sleep for seven hours at night are far better off than those who are allowed to sleep for only three hours. The rule of thumb is: the longer the sleep period, the better. The same rule applies to naps. Naps of 40 minutes in length are certainly better than naps of only ten minutes. Nap durations from 40 minutes to two hours in length are frequently recommended. A variety of laboratory studies (and some actual aviation-performance studies) have shown that naps of these durations are very beneficial.

Sleep Inertia is an Important Consideration

The recommendation that naps be kept to less than 45 minutes or that they be at least two hours in length stems in large part from efforts to minimize sleep inertia (post-nap grogginess). However, we prefer to simply recommend the longest nap allowed by the operational context (although we have no objections to the 45-minute or two-hour standard). It simply should be noted that attempting to control sleep inertia by modifying nap length is probably non-productive in aviation settings.

The logic behind the usual suggestions of nap length (intended to avoid sleep inertia) rarely applies to most pilots because these recommendations are based on what is known about the *typical* sleep architecture of the *average* person who is operating in a *constant schedule* and has been getting *adequate pre-mission sleep*. The 45-minute and two-hour napping recommendation is based on the fact that the theoretical person described above would not be predicted to enter deep sleep until after about 30–45 minutes, and if allowed to sleep more than 100 minutes, he/she then would be expected to have cycled out of deep sleep by the time of awakening. Since sleep inertia is known to be most problematic after arousal from deep sleep, it makes sense to avoid nap lengths that would likely cause awakenings during this phase. Unfortunately, the logic of all of this fails to appreciate the circadian disruptions, pre-duty sleep deprivation, and other factors that tend to increase the homeostatic drive for sleep in a large percentage of civilian and military pilots. Research has shown that the presence of high sleep pressure and certain circadian disruptions can significantly hasten the onset of slow-wave sleep, and exact predictions about the individual manifestations of such a change are impossible to make with any degree of certainty. For instance, studies we have conducted with military pilots have shown that two days of sleep deprivation cause most of them to reach the deepest stage of sleep much more rapidly than normal (usually in less than ten minutes), and to remain in deep sleep far longer than normally observed in well-rested people. So, making generalizations about sleep stages is difficult in the absence of detailed information on the specific individual. Furthermore, it is well known that the placement of naps at certain times in the body's circadian cycle will affect the sleep structure of the nap, making it impossible to predict how long a particular individual will take to reach deep sleep, and once there, how long it will take before a lighter stage of sleep is attained.

Thus, given the difficulties involved in predicting sleep architecture and consequent post-sleep grogginess, it makes more sense to simply assume that sleep inertia will be present upon awakening, and to therefore provide anyone who has been napping with a "grace period" of approximately 20 to 30 minutes to become fully awake before expecting them to handle any type of demanding mental task. Moving around and getting some exercise after awakening may help to shake off the post-sleep sluggishness, so a little physical activity is worth a try. The bottom line is that in the long term, more sleep equals better overall performance, so the idea of restricting nap sleep for the sake of avoiding possible problems with sleep inertia is counterproductive in most situations. Get as much sleep as you can whenever possible.

Stimulant Medications

It has often been said that "drugs and flying just don't mix," and this is certainly true in civil aviation where day-to-day medical supervision is impossible. In fact, the FAA does not authorize flight-crew use of any stimulant other than caffeine (although some herbal supplements, while not specifically sanctioned, are not expressly forbidden). The logic behind this sort of legislation is that problematic levels of aircrew fatigue should be adequately prevented by appropriate fatigue risk management practices. Also in cases where these measures are not 100 percent successful, any commercial crewmember technically has the option of refusing to fly if he/she feels that fatigue has created an unsafe situation. In the event that a civilian pilot declines to take a flight, it might irritate the company and upset the passengers, but no one will be injured or killed as a result. Of course, the situation is different in the military arena (and in civilian emergency settings) where sustained aviation operations are essential to the national defense strategy or to protect people from natural disasters or other unavoidable, life-threatening circumstances. While military pilots, firefighters, and Emergency Medical Service pilots have the same option of refusing to accept flight missions when they feel their performance is impaired, such a decision (to protect the aircraft and the crew by declining a flight) may have the drawbacks of withholding assistance from the innocent people that the crew is sworn to defend, denying the possibility of medical evacuation to injured comrades or civilians, or missing an important opportunity to remove an enemy or other type of threat. For these situations, chemical methods for sustaining alertness may be the better choice.

No matter what the prevailing thoughts are about drugs such as caffeine, modafinil, and amphetamine, the scientific literature is quite clear on at least one point – all are useful for sustaining wakefulness in people who have been deprived of sleep. Only caffeine can be used in civil aviation contexts, but in the military, modafinil and amphetamine may sometimes be authorized. Each alertness-promoting compound has both disadvantages and advantages related to effectiveness, availability, abuse potential, and side effects; but in some settings, none should be ruled out in the absence of careful consideration.

Caffeine

Caffeine's primary advantages are that (1) it is widely available (without a prescription); and (2) it is effective for attenuating sleepiness, especially in people who do not normally consume large quantities of caffeine on a day-to-day basis. Caffeine is found in a variety of products that are easily obtained from the corner store, vending machines, and the galleys of commercial aircraft. Every day people consume caffeine in all sorts of products. Everyone knows about the caffeine in coffee (100–175 mg per cup), but of course there is caffeine in Coke® (31 mg), Mountain Dew® (55 mg), and tea (about 40 mg), as well as energy drinks (Red Bull® has 75–80 mg and 5-Hour Energy® shot has about 200 mg). By the way, these are the amounts of caffeine found in single servings that are typically smaller than what most people consume. The extra-large cup of Starbucks™ can have twice as much caffeine as a typical small cup of home-brewed coffee! Many over-the-counter medications also contain caffeine. For instance, some varieties of Excedrin® have 65 mg of caffeine per tablet, and caffeine may be found in certain cold remedies. Of course, caffeine is also available in preparations such as NoDoz® and Vivarin® (200 mg per pill) as well as in caffeinated candies and chewing gum.

The minimum amount of caffeine recommended to sustain alertness in sleep-deprived people is 200 mg, which in long-term periods devoid of sleep opportunities should be taken approximately every two hours (up to five hours prior to the next sleep break). Generally speaking, a daily dosage limit of 800 mg should not be exceeded. The graph in Figure 12.2 indicates the effects of various single doses of caffeine on the alertness of people who have already been sleep-deprived for two days. This graph shows only

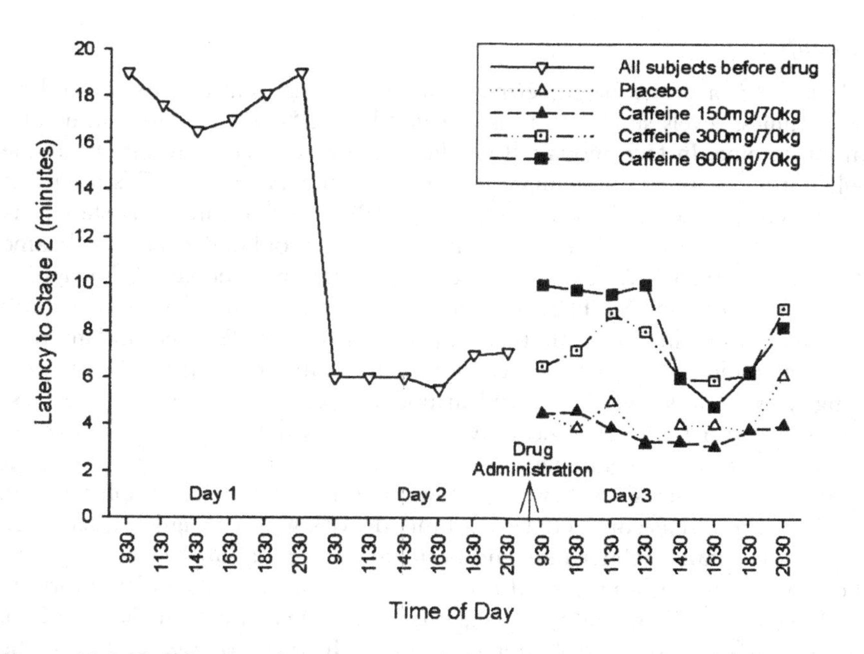

Figure 12.2 Caffeine, in a dose-dependent fashion, has been shown to improve alertness after an extended period of wakefulness

Source: Reprinted with permission from Kamimori, G.H., Johnson, D., Thorne, D., and Belenky, G. (2005), 'Multiple Caffeine Doses Maintain Vigilance During Early Morning Operations,' *Aviation, Space, and Environmental Medicine*, Vol. 76(11), pp. 1046–1050.

sleepiness levels (which obviously did not recover to normal even with the largest dose of caffeine). However, some aspects of performance returned to well-rested levels with the single 600-mg dose.

People who intend to rely on caffeine as a method to postpone sleep should be aware that the body may quickly adjust to the effects of daily caffeine consumption, so heavy caffeine users will probably not get the boost they really need from those two cups of coffee when they are fighting off sleep in the middle of the night. Those who find they frequently need some help staying alert on night shift or when flying those multi-leg long-haul missions (or those seemingly interminable 18–40 hour military sorties) should only use caffeine during the times when they really need it. Although it is sacrilege to make such a suggestion, these people should switch to decaffeinated products on days when fatigue is not expected to be a significant problem. However, when operational demands make pre-mission sleep difficult or impossible to obtain, and a long stint of duty is on its way, caffeine should be considered a "first-line" pharmacological approach to sustaining alertness and performance in sleepy individuals.

Modafinil

Modafinil is a prescription alertness-enhancing drug that appears to produce wakefulness levels similar to those created by the "gold-standard" stimulant, amphetamine. In fact, modafinil has become the alternative to amphetamine when it is necessary to attenuate sleepiness in military settings. This drug was approved for use in the US in December 1998, and research has shown its safety and effectiveness in operational settings. It is sold under the brand name Provigil® in the US. An extended-release formulation, armodafinil (Nuvigil®), was approved by the FDA in 2007. Researchers have found that both formulations maintain the alertness of people with sleep disorders and improve the functioning of people who cannot sleep because of night work or really long duty periods. Modafinil and armodafinil effectively maintain alertness without causing the rapid heartbeats and increased blood pressure that are associated with amphetamines. Also, both can be used with less rigid controls than amphetamine (since they have a lower abuse potential than amphetamine).

The military has conducted some limited work with Provigil®; it has been found to sustain the alertness and performance of UH-60 helicopter pilots, keeping them working at well-rested levels, even at 05:00 in the morning and even after 22 straight hours without sleep (as shown in Figure 12.3). Scientists from both the Walter Reed Army Institute of Research and the Air Force Research Laboratory have performed modafinil studies (one of the Air Force studies was performed on F-117 pilots), and both groups have reported positive results with few, if any, problems.

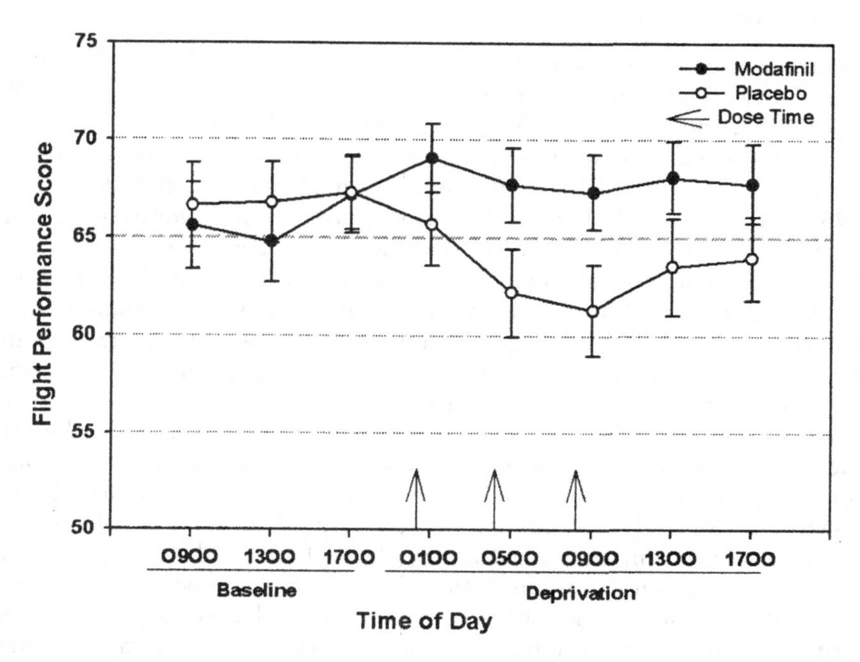

Figure 12.3 Modafinil (Provigil®) is effective for short-term fatigue management
Source: Caldwell, J.A., Smythe, N.K., Caldwell, J.L., Hall, K.K., Norman, D.N., Prazinko, B.F., Estrada, A., Johnson, P.A., Crowley, J.S., and Brock, M.E. (1999), *The Effects of Modafinil on Aviator Performance During 40 Hours of Continuous Wakefulness: A UH-60 Helicopter Simulator Study*, USAARL Report No. 99–17 (Public Domain).

The recommended dose of modafinil is either 100 or 200 mgs every eight hours during long-term sleep deprivation (that is, 40 hours without sleep), not to exceed 400 mgs in any given 24-hour period. As is the case with any alertness-enhancing drug, the smaller dose is more likely to be effective in situations where the amount of sleep deprivation is relatively small, or in situations where the drug is administered prior to the appearance of fatigue-related problems. Once again, it is always better to *prevent* performance decrements than to attempt to *restore* performance that has already begun to degrade.

A couple of final points about medication use in military flight operations is that (1) each aircrew member should first be given a test dose during a controlled, non-flying period to safeguard against any unlikely idiosyncratic reactions; and (2) prescription medications should only be used when proper medical oversight is feasible.

Dextroamphetamine

Dextroamphetamine has been the most heavily researched prescription stimulant available today. Amphetamines have been on the market in the US since 1937, and these drugs have been widely used to treat the symptoms of medical conditions such as narcolepsy (a disorder of excessive daytime sleepiness) and hyperactivity/attention deficit disorder. In the 1940s and 1950s, the military found amphetamines were effective for restoring or maintaining the performance of sleep-deprived people to non-deprived levels. Although very valuable, these compounds must be used carefully to minimize the possibility of addiction and/or abuse. However, it is noteworthy that the US Air Force has successfully used amphetamines for years without any evidence of aircrew addiction or other problems affecting the flight status of the pilots involved. In addition, there has never been an Air Force aviation mishap in which dextroamphetamine was found to be a contributing factor whereas untreated fatigue has been at least partially responsible for numerous accidents. While modafinil has generally become the first choice when a pharmacological countermeasure is needed, dextroamphetamine (Dexedrine®) is still authorized under US Air Force and Army policy for certain situations today because the effects of this compound have been extensively studied in the laboratory and in the field. Research at the US Army Aeromedical Research Laboratory determined that multiple 10 mg doses of dextroamphetamine, administered prior to the onset of fatigue degradations, sustained the performance of helicopter pilots throughout 60 hours of continuous wakefulness without producing unwanted side effects (although the medication did increase blood pressure, and several pilots said they could "feel their heart beating faster"). Interestingly, no convincing support has been found for the widely held opinion that amphetamines (used in the fashion authorized by the US military) make people reckless and overconfident, whereas there is plenty of evidence that these medications reverse the increased carelessness that often accompanies fatigue. The effects of dextroamphetamine on the simulator flight performance of severely fatigued pilots are shown in Figure 12.4.

Field studies of Dexedrine® in aviation also have been favorable. Dexedrine® was given to EF-111A Raven jet crews during an electronic jamming mission associated with the Air Force strike on Libya in April of 1986, and the drug enabled crews to overcome the fatigue of the mission itself and the sleep deprivation which occurred during earlier preparation for the mission, without producing in-flight or landing problems. Dexedrine® also was given to F-15C pilots flying lengthy combat air patrol missions during Operation Desert Shield/Storm, and it was found that the stimulant enabled flight crews to overcome fatigue from sleep deprivation and circadian disruptions. No adverse effects were reported, and no aviators expressed a need to continue the drug once proper work/sleep schedules were reinstated.

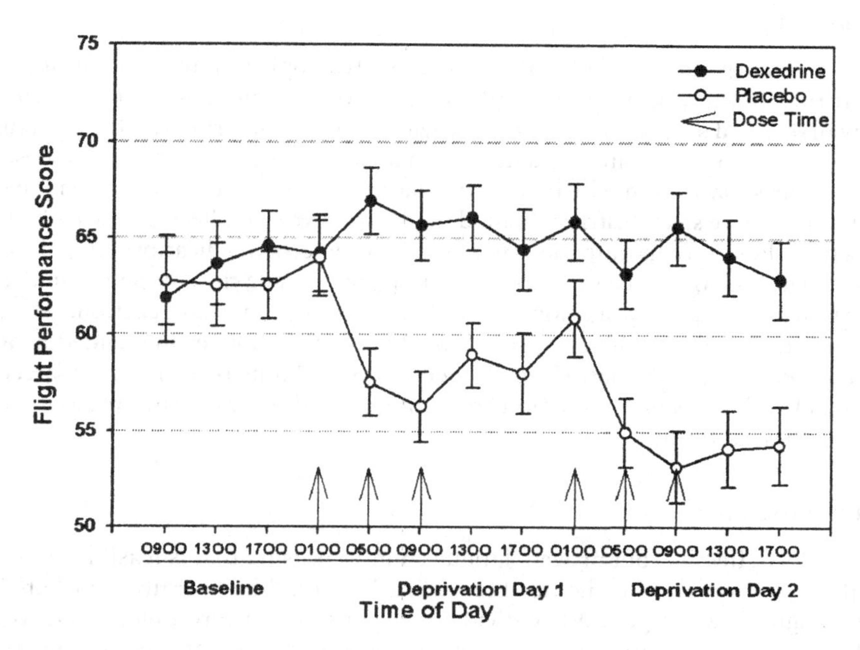

Figure 12.4 Dextroamphetamine has proven useful in military aviation

Source: Reprinted with permission from Caldwell, J.A., Smyth, N.K., LeDuc, P., and Caldwell, J.L. (2000), 'Efficacy of Dexedrine for Maintaining Aviator Performance During 64 Hours of Sustained Wakefulness: A Simulator Study,' *Aviation, Space, and Environmental Medicine*, Vol. 71, pp. 7–18.

This agrees with the results of a large survey of Air Force pilots (at the conclusion of the Gulf War), which indicated that dextroamphetamine was helpful for maintaining mission performance during sustained operations without inducing unwanted side effects. Thus, while Dexedrine® is not available to civilian pilots, military aviators at times may be authorized to use this medication as a last resort in situations where adequate sleep is impossible to obtain.

A recommended dosing schedule for dextroamphetamine is 10 mg at midnight, 04:00, and 08:00 (or similar times) in situations involving a full night of sleep loss followed by a subsequent requirement to spend the next day on the job. A pre-mission test-dose experience (with only 5 or 10 mg) prior to the actual military context in which the drug will be used to combat the effects of unavoidable episodes of sleep loss is required by anyone allowed to use this medication in operational contexts. Also, as noted earlier, medical oversight is a must.

Summary

In summary, caffeine, modafinil, and dextroamphetamine can all stave off the effects of fatigue, but caffeine is the only compound that is widely available and socially acceptable. Caffeine is an option primarily for civilian aviators who are not involved in combat aviation operations. In very intense wartime situations in which military aviators are required to continue flying despite severe sleep restriction, modafinil and dextroamphetamine are likely better choices than caffeine. However, any type of medication should be taken at least once in a safe environment prior to using the compound under actual operational conditions as a check for any unusual reaction. Also, remember that the chronic use of any stimulant is not recommended as a replacement for adequate sleep. Proper crew scheduling is essential for safety and effectiveness in the aviation environment, and the only true antidote for sleepiness is sleep!

Rest Breaks

A mildly effective, but proven, fatigue countermeasure that is feasible in just about any situation is the rest break. Breaks should be liberally distributed throughout work periods, particularly when things are routine, repetitive, long, and/or monotonous. It has long been established that breaks improve performance by allowing physiological recovery, reducing boredom, and increasing worker satisfaction. In industrial settings, the Hawthorne experiments indicated that job performance and productivity improved when work breaks were introduced. In aviation contexts, the same logic applies. Dr David Neri performed a study in the 747–400 simulator at NASA Ames Research Center to determine the effects of hourly breaks on the alertness of pilots flying six-hour night flights. As shown in Figure 12.5, he found that brief (seven minute) hourly opportunities to get up, walk around, and converse with other crewmembers during the flight reduced physiological sleepiness (as measured by electroencephalography) and subjective sleepiness (as measured by self-reports) for at least 15 to 25 minutes around the time of the circadian trough. Although the effects of the breaks were short lived, they could be extremely beneficial if the breaks are placed right before critical periods in the flight.

Military pilots likewise can be expected to benefit from rest breaks. Research performed by Drs Angus, Pigeau, and Heslegrave established that breaks were helpful for temporarily overcoming the fatigue associated with sustained operations. Thus, while optimum schedules for work breaks have not been determined, it is clear that this fatigue countermeasure works and is feasible. In fact, the NSF currently recommends that drivers schedule rest breaks every two hours or 100 miles on long trips in order to attenuate the effects of driver fatigue.

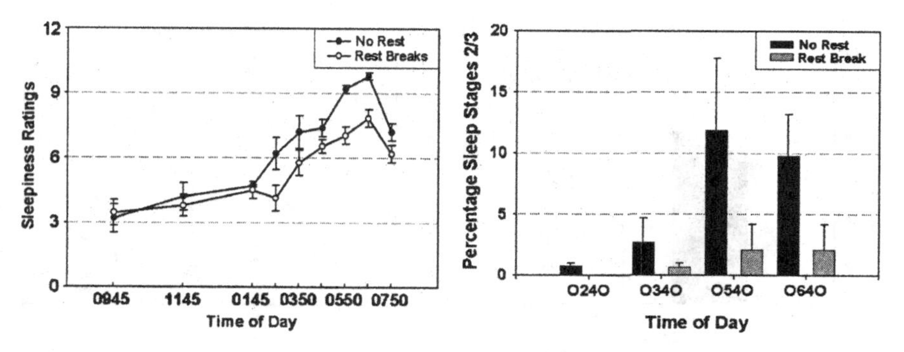

Figure 12.5 Short breaks can temporarily restore performance

Source: Reprinted with permission from Neri, D.F., Oyung, R.L., Colletti, L.M., Mallis, M.M., Tam, P.Y., and Dinges, D.F. (2002), 'Controlled Breaks as a Fatigue Countermeasure on the Flight Deck,' *Aviation, Space, and Environmental Medicine*, Vol. 73, pp. 654–664.

Exercise

Physical exercise is often suggested as a way to stave off the effects of fatigue in situations where naps or other sleep periods are not possible. Unfortunately, this strategy is not particularly useful in the in-flight environment for obvious reasons. However, for pilots engaged in short-hop operations, or those stuck on the ground to handle administrative work, exercise should be considered when fatigue is reaching problematic levels. Although some of the effects of exercise may be related simply to the fact that bouts of physical activity actually constitute a break from the routine of work, there is evidence that exercise may be an effective, even though temporary, method for increasing alertness and arousal. Researchers have discovered that moderate physical exercise improves the cognitive performance of non-sleep-deprived volunteers, and that fairly stressful exercise enhances cognitive vigilance in sleep-deprived people. Researchers at the US Army Aeromedical Research Laboratory found that sleep-deprived pilots who engaged in hourly ten-minute bouts of moderately difficult physical activity (running on a treadmill) experienced greater physiological arousal and less sleepiness for several minutes after each exercise cycle, as shown in Figure 12.6. Thus, like rest breaks, physical exercise effectively promotes alertness in the short term, but the effects are relatively short lived (lasting only 10 to 30 minutes). In situations where frequent exercise breaks are feasible, the use of this countermeasure may be appropriate; however, exercise is generally not practical in most aviation environments. Also, it should be noted that even in situations where exercise is possible, heavy exercise should be avoided since there is evidence that intense physical exertion ultimately makes sleepiness more pronounced after its initial benefits subside.

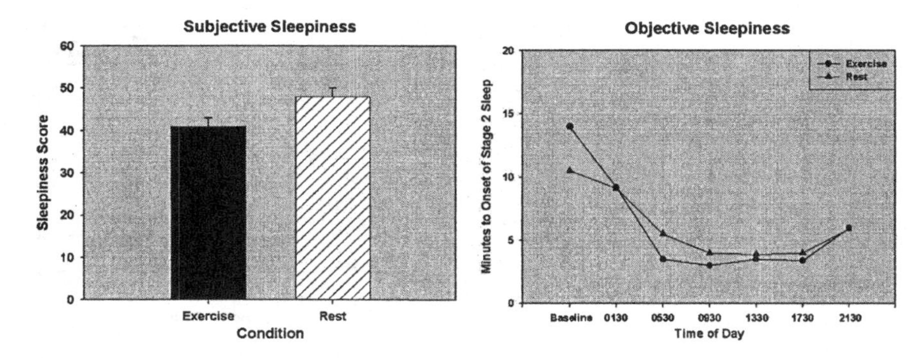

Figure 12.6 Physical exercise can briefly sustain alertness during prolonged wakefulness periods

Source: From LeDuc, P.A., Caldwell, J.A., Ruyak, P.S., Prazinko, B., Gardner, S., Colon, J., Norman, D., Cruz, V., Jones, R., and Brock, M. (1998), *The Effects of Exercise as a Countermeasure for Fatigue in Sleep Deprived Aviators*, USAARL Report No. 98–35 (Public Domain).

Postural Changes

Although few studies have examined the influence of body posture on the alertness of individuals who are fatigued, some investigators have found that it is easier to maintain wakefulness when sitting or standing than when lying down. Also, it has been established that people are more likely to stay awake when seated upright than when reclining. Shift workers who suffer from drowsiness on the job often state that standing up is one way they improve their alertness at work, and patients who suffer from narcolepsy have anecdotally reported that their ability to stand up at school or at work was essential to their success in college and on the job. Why this works is not completely understood, but it has been suggested that the alertness-enhancing benefits of upright posture may stem from changes in sensory input. Of course, it is also true that postural changes are often associated with brief activity breaks, and in some settings, this may partially explain the beneficial effects of moving to a more upright position. We have determined that sleep-deprived pilots are more alert and more vigilant when they are allowed to stand than when they are required to sit. As shown in Figure 12.7, the positive effects of the more upright posture were seen both in terms of brain activity patterns and in terms of the pilots' abilities to monitor and respond to a monotonous reaction time task. Of course, it is not a good idea for pilots to try to stand up while flying, but they should avoid slouching down in their seats or otherwise assuming a reclined position when they are fighting the effects of fatigue. Also, when engaged in ground-based operations, they should take advantage of the beneficial

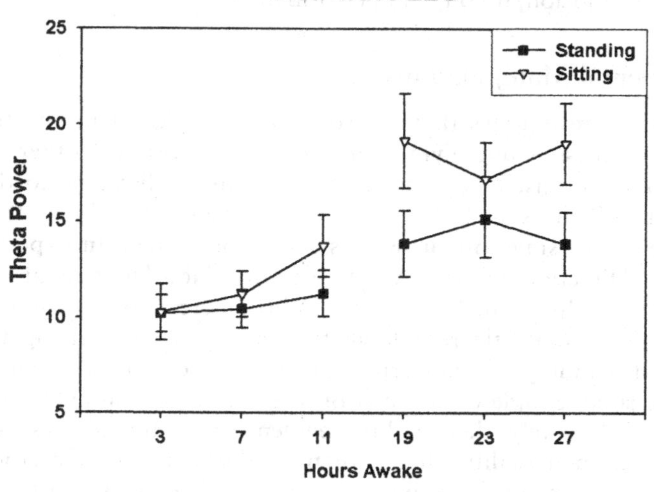

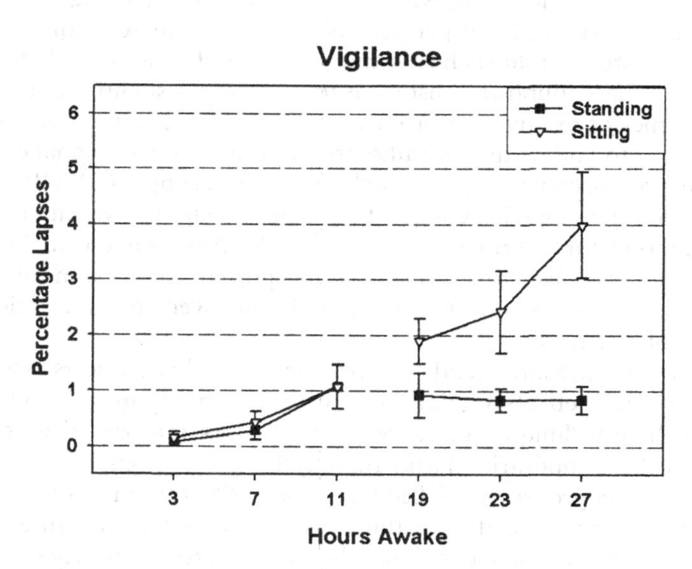

Figure 12.7 Maintaining an upright posture can increase alertness levels during times when fatigue is strong

Source: Reprinted with permission from Caldwell, J.A., Prazinko, B., and Caldwell, J.L. (2003), 'Body Posture Affects Electroencephalographic Activity and Psychomotor Vigilance Task Performance in Sleep-Deprived Subjects,' *Clinical Neurophysiology*, Vol. 114, pp. 23–31.

effects of maintaining an upright posture any time sleepiness threatens to impair concentration, mood, and performance.

Environmental Manipulations

Two common techniques that are sometimes employed in an attempt to improve alertness while flying, driving, or performing other types of tasks involve increased exposure to cold air and/or listening to the radio. Sleep-deprived pilots will often aim the air vents directly at their faces in an attempt to stave off drowsiness, and sometimes, an expertly tuned Automatic Direction Finder (ADF) (used for Non-Directional Beacon – NDB – approaches) can bring some radio entertainment to the cockpit. Sleepy drivers frequently roll down the windows and turn up the radio in hopes of avoiding an inadvertent sleep attack on the way home late at night, and some people even pinch or slap themselves in an effort to stay awake! Unfortunately, there is little evidence that any of these strategies are effective. In actuality, there are no published scientific studies that document any benefits from exposure to cold air or listening to the radio on the maintenance of flight performance. However, there is some evidence from driving studies that such environmental modifications can help a little. Some researchers found that listening to a car radio slightly improved the reaction times of extroverted or inexperienced drivers, but others reported that listening to audio tapes or radio programs had only marginal effects on the driving performance of partially sleep-deprived subjects. Drs Reyner and Horne found that applying cold air to the faces of drivers was an ineffective technique to restore alertness (see Figure 12.8). Also, while listening to the radio temporarily decreased ratings of self-reported sleepiness more than the cold air, the effect was not associated with improvements in physiological measures of alertness.

Such countermeasures tend to have very short-lived effects when they have any positive effects at all. Primarily, the benefits seem to stem from the fact that these techniques introduce some novelty into the work environment, and this temporarily masks the physiological fatigue that is present. However, the improvement is fleeting at best. Thus, the use of such strategies is not recommended. If nothing else is available, you can give these techniques a try, but the focus should be to switch to a more effective countermeasure as soon as possible.

Physical Fitness

As a group, aviators tend to be physically fit because of their outgoing, active personalities as well as the requirements to pass frequent flight physicals. Clearly, physical fitness is important to the overall quality of life, and it is known that people who are healthy and fit tend to sleep better than people who are overweight and out of shape. But does increased physical

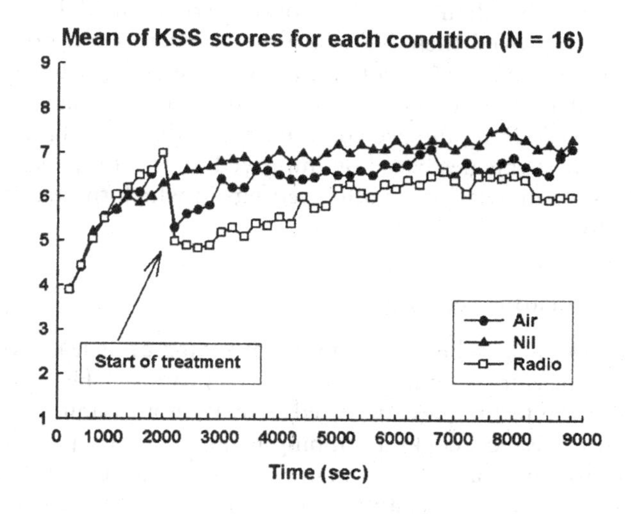

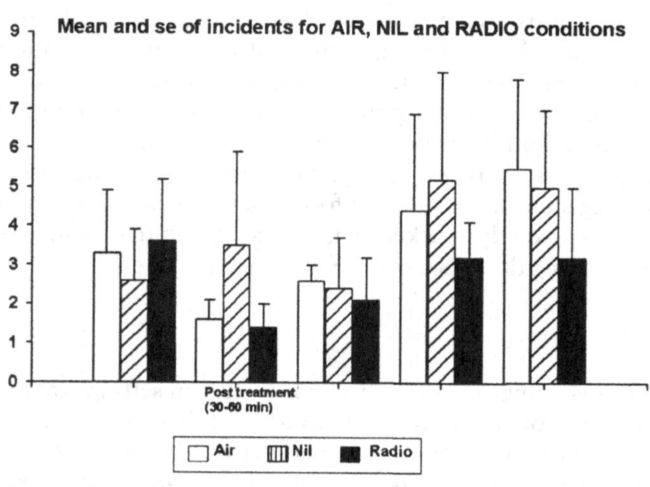

Figure 12.8 Sole dependence on environmental strategies is not advisable

Source: Reprinted with permission from Reyner, L.A., and Horne, J.A. (1998), 'Evaluation of "In-Car" Countermeasures to Sleepiness: Cold Air and Radio,' *Sleep*, Vol. 21, pp. 46–51.

fitness reduce the vulnerability to the effects of fatigue? There seems at least to be an assumption that physical fitness can reduce the impact of sustained work on personnel, because the military has used this approach for years. Unfortunately, while it is true that fit people are better able to withstand longer periods of physical activity, there is little evidence that the same applies to the performance of cognitive tasks. Dr Härmä and colleagues found that female night workers complained less of fatigue after they had improved their physical fitness, but this did not appear to translate into performance benefits on a cognitive task. Other researchers have found that highly fit individuals are no better off than their less fit compatriots in sustaining intense cognitive work during sleep deprivation. Thus, anti-fatigue strategies based on improved physical fitness are unlikely to produce results in aviation or other settings in which the work is mental rather than physical.

Generally Applicable Techniques

Hopefully, everyone will be able to take advantage of at least some of the fatigue countermeasures that have already been proposed. However, in some cases, many of the techniques suggested here are not permitted in aviation operations. For instance, cockpit napping is not currently permitted under FAA regulations, and commercial aviators often cannot take full advantage of rest breaks or postural changes simply because they are required to remain in their seats except for "biological necessities" which, strangely enough, sometimes does not include either sleep or the need to move about in order to become more alert. However, almost everyone can engage in stimulating conversation to keep the brain active, rotate various flight duties to minimize boredom, squirm around in the seat and stretch to enhance physical arousal, and avoid sugary, fat-laden foods that may result in feelings of sluggishness. Protein-rich foods may be beneficial. Granted, these strategies are not exactly "rocket fuel," but anything that reduces sleepiness is worth trying. Just don't bet your life on such techniques!

Summary

In summary, it is a huge challenge to fight fatigue in settings that are devoid of adequate sleep opportunities. Obviously, a full seven to eight hours of pre-mission sleep is the best way to prevent on-the-job fatigue every day, but if this is not possible, napping is an excellent standby. If neither of these strategies is feasible, there are other countermeasures that have been proven variably successful in different types of situations. Just make sure that the one you choose is backed up by solid scientific research and employed in a manner that has been shown effective in the past.

Top Ten Points About Anti-fatigue Strategies for Sleep Restriction

- Settings where sleep opportunities are limited or non-existent pose special problems with fatigue.
- Use strategic napping to help satisfy the body's physiological drive for consolidated sleep.
- Be sure to follow specific scientific guidance when establishing napping countermeasures.
- In civilian and military aviation, rely on caffeine to temporarily minimize the effects of sleep deprivation.
- In military sustained operations, if authorized, use modafinil or dextroamphetamine to maintain alertness when sleep opportunities are non-existent.
- When feasible, implement brief and frequent rest breaks to bolster alertness.
- In ground-based operations, try short but non-strenuous periods of exercise to help stay awake.
- In ground-based operations, work while standing up rather than while sitting down when possible.
- Rotate duties, engage in conversation, change the temperature, change the lighting, move around in the seat, and do anything else you can think of to minimize the effects of boredom when sleepy.
- Stay in good physical condition, but don't count on this alone to make yourself fatigue resistant.

13 Issues to Consider When Launching a Program of Fatigue Management

By now everyone should be convinced (1) that fatigue can be a real problem in aviation and other transportation settings; (2) that impaired alertness stems from genuine physical/biological limitations that cannot be overcome through training or motivational incentives; and (3) that there are in fact scientifically valid methods for managing alertness in both civil and military aviation. Although fatigue management was long considered to be a function of limiting time on duty, in truth, the bottom line is that most of the "fatigue problem" could be resolved simply by ensuring that everyone gets enough sleep and sticks to a somewhat consistent schedule. But, as we all know, there is a host of societal, economic, and attitudinal barriers to creating a world in which sleep receives the attention it deserves in operational settings. Interestingly enough, several years ago it became apparent that formal fatigue-management policies were not even being used in many of the sleep-disorders centers and sleep-research laboratories that were providing information about the seriousness of the problem! How can commercial airlines and military units be expected to correctly address the issue of fatigue if the experts are not doing it in their own organizations?

As it turns out, the sleep experts are increasingly taking their role-model status more seriously, and as a result, well-designed alertness-management programs are being implemented throughout the clinical and research sleep communities. Meanwhile, everyone else seems to be learning from their past mistakes, and this is raising the "fatigue-awareness IQ" of the general public, corporations, the aviation industry, the military, and others. In fact, the National Consumer Research Institute recently included sleep improvement among their top five predicted health trends for 2012. According to the Values Institute, 76 percent of Americans now want to improve the quantity and quality of their sleep – a statistic that is quite encouraging in light of the fact that the World Association of Sleep Medicine says sleep deprivation affects the quality of life of 45 percent of the world's population.

The American College of Occupational and Environmental Medicine (ACOEM) published a position paper in 2012 that emphasized the importance of fatigue management for any organization concerned with the health, safety, and productivity of its employees, especially for safety-sensitive operations such as healthcare, energy, and transportation. The ACOEM states that fatigue management efforts within safety management systems (SMS) should go far beyond the reactive practices of the past to include an "explicit and comprehensive process for measuring, mitigating, and managing the fatigue risk to which a company is exposed," and that this process should be integrated into and supported by an overarching SMS. Doing so creates a FRMS that contains the features designed and outlined by Dr Moore-Ede in 2009. Specifically, the key concepts required for successful FRMS implementation are as follows:

1. It must be *science-based* – supported by established peer-reviewed scientific publications.
2. It should be *data-driven* – decisions are not based on opinions, but rather on the collection and objective analysis of data.
3. It must be *cooperatively designed* by all stakeholders working together.
4. It must be *fully implemented* across the entire organization to ensure system-wide use of tools, systems, policies, and procedures.
5. It should be *integrated* into the corporate safety and health management systems.
6. It should not remain a static entity, but rather, it should be *continuously improved* to progressively reduce risk using feedback, evaluation, and modification.
7. It should be proactively *budgeted* and *justified* by an accurate return on investment (ROI) business case.
8. It must be *owned* and accepted by the senior corporate leadership as a priority responsibility.

Although the concept of FRMS is relatively new, it is rapidly being adopted throughout the transportation industry. The Airline Safety and Federal Aviation Administration Extension Act of 2010 requires US airlines to rapidly establish comprehensive Fatigue Risk Management Plans (FRMPs), and there are provisions for airlines to implement FRMS-based fatigue-management alternatives based on science, data, and monitoring of operations that exceed published flight-scheduling requirments. Other organizations, to include mining, oil and gas, and healthcare, are taking similar steps.

Organizations that choose not to establish proactive fatigue-management plans are placing themselves at increased risk of being found criminally negligent in the event of an on-the-job mishap as well as in the event of an injury or death that results from a fall-asleep crash during the drive home from work. It is now unacceptable for organizations to justify inaction by claiming ignorance of the problem. Although a direct causal relationship between sleep deprivation and accidents is sometimes difficult to establish, it is well known that sleep loss is associated with a variety of decrements in performance, judgment, cognition, and mood which place personnel at risk of serious errors with potentially disastrous consequences. This information is being used to hold organizations responsible for the mistakes of their fatigued employees both on the job and off.

Putting a formal fatigue management plan in place shows that the risks resulting from sleepy personnel are known and that the organization has chosen to actively counter these risks – both on the job and off. While it may not be easy to change personal habits, cultural biases, and traditional industrial practices, it is not impossible, and a high-quality FRMS design in fact addresses human concerns as well as a variety of system issues. In a good FRMS, key defenses against fatigue are integrated into an overall program that ensures employees are getting sufficient sleep, that they are free from fatigue-related issues due to factors such as sleep disorders, that controls are in place to minimize the impact of any fatigue-related errors that do in fact occur, and that the controls are periodically assessed to ensure their effectiveness.

In line with Dr Moore-Ede's recommendations, the ACOEM points out that it is important for any FRMS to include several key defenses against fatigue-related incidents and accidents. These are:

1. Ensuring adequate staffing levels and workload balance to prevent a worsening of the many unavoidable fatigue-related problems associated with shift work.
2. Minimizing to the extent possible schedule-related fatigue factors by utilizing biomathematical models to identify the risks associated with specific work/rest schedules, guide the implementation of fatigue countermeasures, aid in accident investigations, and reinforce counter-fatigue educational efforts.
3. Educating employees about fatigue-related work and social/familial hazards; the importance of sleep, circadian rhythms, and lifestyle factors in the fatigue equation; how to obtain adequate sleep; how to recognize and access treatment for sleep disorders; and how to effectively utilize valid alertness-management strategies. Note that management plays a key role in not only supplying information, but in providing the motivation and sometimes the resources necessary for employees to report to work in a well-rested state.
4. Ensuring the workplace environment promotes alertness in terms of adequate and proper lighting, humidity and noise control, and ergonomic design. In addition, the impact of the type and duration of work on fatigue as well as the importance of naps, rest breaks, and balanced nutrition in fatigue management should be emphasized.
5. Making certain that employees, coworkers, and supervisors are able to quickly recognize the signs of excess fatigue and that there are provisions for actions to immediately mitigate either the fatigue itself or the

risks posed by fatigue. Actions may include switching the employee to a less safety-sensitive role, augmenting peer-based double-checking procedures, using caffeine to temporarily boost alertness, or changing the type or intensity of environmental lighting.

These defenses against fatigue should be a part of recognized company procedures, fully backed by management, and supported by employees who are aware that fatigue management is a shared responsibility. And once the FRMS is implemented, it must be continuously evaluated and improved based on data derived through investigation of any fatigue-related incidents and accidents, assessment of productivity levels and absenteeism patterns, evaluation of accident/injury rates, and tracking of health-related costs. A corporate commitment to building and maintaining an effective FRMS includes setting key fatigue-management goals, strategies to measure progress toward those goals, provisions for periodic gap analyses, and workable plans to close any gaps between desired and actual performance indicators.

Of course, institutionalized alertness-management strategies can only be successful if everyone who makes up the company, the unit, or the shift accepts his or her own personal responsibilities to alleviate fatigue both on and off the job. The best organizational guidelines in the world will have little effect unless everyone supports the overall plan and agrees to follow the guidance provided. Remember, individual action is the key to any kind of organizational success.

So, do your part to improve alertness in the aviation environment. Support the development of effective fatigue management plans and policies. Contribute to educational efforts regarding the causes of fatigue, the potential consequences of drowsiness on the job, and the proactive steps that can be taken to maximize alertness on the ground and in the air. Emphasize the benefits of fatigue management in terms of improved productivity and safety, higher employee satisfaction and staff retention, better personal well-being, and enhanced crew functioning. Set a good example through your own actions. Your efforts will go a long way toward making sure that you and your crew are always *awake at the stick*!

Top Ten Points to Consider when Launching a Fatigue-Management Program

- It is time for individuals and organizations to recognize the impact of fatigue and to commit to doing something about it.
- A data-driven approach to fatigue risk management is the key to safety success.
- Leaders must take steps to ensure personnel get the sleep they need, and personnel must make sleep a personal priority.
- Everyone should be educated about sleep and fatigue via company-supported and web-based training efforts.
- Corporations should optimize schedule design and adjust staffing ratios to help control the problems stemming from shift lag and jet lag.
- Companies should systematically implement fatigue countermeasures to enhance individual safety, well-being, and performance.
- Employees and management must work cooperatively to maximize off-duty rest, minimize circadian disruptions, and optimize on-duty performance.
- Employees and management must work together to seek sound solutions to fatigue-related problems through consults with aviation fatigue experts and/or sleep specialists.
- Organizations and individuals must realize that fatigue-related mishaps on duty and off duty may result in serious litigation and financial loss.
- Proactive individual and corporate efforts will go a long way toward ensuring that aviators are fully awake at the stick.

References and Suggested Reading

Airbus (2015), 'The A340–500 – The Ultra Long-Range Machine, www.airbus.com/aircraftfamilies/ (accessed 29 December 2015).

Akerstedt, T. (1995a), 'Work Hours, Sleepiness and the Underlying Mechanisms,' *Journal of Sleep Research*, Vol. 4(Suppl. 2), pp. 15–22.

Akerstedt, T. (1995b), 'Work Hours and Sleepiness,' *Neurophysiology Clinical*, Vol. 25(1), pp. 367–375.

Akerstedt, T. (1988), 'Sleepiness as a Consequence of Shift Work,' *Sleep*, Vol. 11(1), pp. 17–34.

Akerstedt, T., and Gillberg, M. (1990), 'Subjective and Objective Sleepiness in the Active Individual,' *International Journal of Neuroscience*, Vol. 52(1–1), pp. 29–37.

American College of Occupational and Environmental Medicine (2012), 'Fatigue Risk Management in the Workplace,' *Journal of Occupational and Environmental Medicine*, Vol. 54(2), pp. 231–258.

Angus, R.G., Pigeau, R.A., and Heslegrave, R.J. (1992), 'Sustained-Operations Studies: From the Field to the Laboratory,' In C. Stampi (Ed.), *Why We Nap*, pp. 217–241, Kirkäuser, Boston, MA.

Balkin, T., Thorne, D., Sing, H., Thomas, M., Redmond, D., Wesensten, N., Williams, J., Hall, S., and Belenky, G. (2000), 'Effects of Sleep Schedules on Commercial Motor Vehicle Driver Performance,' *Department of Transportation Federal Motor Carrier Safety Administration Report No. DOT-MC-00–133*, U.S. Department of Transportation, Washington, DC.

Banks, S., Van Dongen, H.P.A., Maislin, G., and Dinges, D.F. (2010), 'Neurobehavioral Dynamics Following Chronic Sleep Restriction: Dose-response Effects of One Night for Recovery,' *Sleep*, Vol. 33(8), pp.1013–1026.

Battelle. (1998), *An Overview of the Scientific Literature Concerning Fatigue, Sleep, and the Circadian Cycle*, A report prepared for the Office of the Chief Scientific and Technical Advisor for Human Factors Federal Aviation Administration, www.alliedpilots.org/pub/presskit/safety/battellereport.html (accessed 29 December 2015).

Belenky, G., Penetar, D.M, Thorne, D., Popp, K., Leu, J., Thomas, M., Sing, H., Balkin, T., Wesensten, N., and Redmond, D. (1994), 'The Effects of Sleep Deprivation on Performance during Continuous Combat Operations,' In *Institute of Medicine Committee on Military Nutrition Research, Food Components to Enhance Performance*, National Academy Press, Washington, DC.

Belenky, G., Wesensten, N.J., Thorne, D.R., Thomas, M.L., Sing, H.C., Redmond, D.P., Russo, M.B., and Balkin, T.J. (2003), 'Patterns of Performance Degradation and Restoration During Sleep Restriction and Subsequent Recovery: A Sleep Dose-Response Study,' *Journal of Sleep Research*, Vol. 12(1), pp. 1–12.

Berry, R.B., Broks, R., Gamaldo, C.E., Harding, S.M., Marcus, C.L., and Vaughn, B.V. for the American Academy of Sleep Medicine. (2012), *The AASM Manual for the Scoring of Sleep and Associated Events: Rules, terminology, and technical specifications, Version 2.0.* www.aasmnet.org, American Academy of Sleep Medicine, Darien, IL.

Bindley, S. (1997), *Biological Clocks: Your Owner's Manual*, Overseas Publishers Association, Amsterdam.

Boeing (2001a), *Current Market Outlook 2000: In to the New Century, The Demand for Air Travel*, www.boeing.com/commercial/cmo/3at00.html (accessed 29 December 2015).

Boeing (2001b), 'More of a Good Thing: Boeing 777 Longer-Range Derivatives Continue the Family Tradition,' www.boeing.com/commercial/777family/background_lr.html (accessed 29 December 2015).

Bonnet, M.H., Gomez, S., Wirth, O., and Arnad, D.L. (1995), 'The Use of Caffeine versus Prophylactic Naps in Sustained Performance,' *Sleep*, Vol. 18(2), pp. 97–104.

Broughton, R.J., and Ogilvie, R.D. (1992), *Sleep, Arousal, and Performance*, Birkhauser, Boston, MA.

Caldwell, J.A. (1997), 'Fatigue in The Aviation Environment: An Overview of the Causes and Effects as well as Recommended Countermeasures,' *Aviation, Space, and Environmental Medicine*, Vol. 68(10), pp. 932–938.

Caldwell, J.A. (2002), 'Fatigue Facts for Aviators … and Everybody Else!' *Flying Safety*, September, pp. 20–25.

Caldwell, J.A. (2012a), 'Understanding and Managing Fatigue in Aviation,' In G. Matthews, P.A. Desmond, and P.A. Hancock (Eds) *Handbook of Operator Fatigue*, pp. 379–392, Ashgate Publishing Co, Farnham.

Caldwell, J.A. (2012b), 'Sleep Deprivation and Aviation Performance,' *Current Directions in Psychological Science*, Vol. 21(2), pp. 85–89.

Caldwell, J.A., and Caldwell, J.L. (1998), 'Comparison of the Effects of Zolpidem-Induced Prophylactic Naps to Placebo Naps and Forced-Rest Periods in Prolonged Work Schedules,' *Sleep*, Vol. 21(1), pp. 79–90.

Caldwell, J.A., Caldwell, J.L., and Schmidt, R.M. (2008). 'Alertness Management Strategies for Operational Contexts,' *Sleep Medicine Reviews*, Vol. 12(4), pp. 257–273.

Caldwell, J.A., Caldwell, J.L., Smythe, N.K., and Hall, K.K. (2000), 'A Double-Blind, Placebo-Controlled Investigation of the Efficacy of Modafinil for Sustaining the Alertness and Performance of Aviators: A Helicopter Simulator Study,' *Psychopharmacology*, Vol. 150(3), pp. 272–282.

Caldwell, J.A., Gilreath, S.R. and Erickson, B.S. (2002), 'A Survey of Aircrew Fatigue in a Sample of Army Aviation Personnel,' *Aviation, Space, and Environmental Medicine*, Vol. 73(5), pp. 472–480.

Caldwell, J.A., Hall, K.K., and Erickson, B.S. (2002), 'EEG Data Collected from Helicopter Pilots in Flight Are Sufficiently Sensitive to Detect Increased Fatigue from Sleep Deprivation,' *International Journal of Aviation Psychology*, Vol. 12(1), pp. 19–32.

Caldwell, J.A., and Mallis, M.M. (2013), 'Sleep in Aviation and Space,' In C.A. Kushida (Ed.) *The Encyclopedia of Sleep*, Vol. 2, pp. 724–728. Academic Press, Waltham, MA.

Caldwell, J.A., Mallis, M.M., Caldwell, J.L., Paul, M.A., Miller, J.C., and Neri, D.F. (2009), 'Fatigue Countermeasures in Aviation,' *Aviation, Space, and Environmental Medicine*, Vol. 80(1), pp. 29–59.

Caldwell, J.A., Mu, Q., Smith, J.K., Mishory, A., Caldwell, J.L., Peters, G., Brown, D.L., and George, M.S. (2005), 'Are Individual Differences in Fatigue Vulnerability Related to Baseline Differences in Cortical Activation?' *Behavioral Neuroscience*, Vol. 119(3), pp. 694–707.

Caldwell, J.A., Smythe, N.K., LeDuc, P.A., Prazinko, B.F., Caldwell, J.L., Norman, D.N., Skoumbourdis, E., Estrada, A., Sprenger, W.D., Ruyak, P.S., and Hoffman, S. (1999), 'The Efficacy of Dexedrine for the Sustainment of Performance during 64 Hours of Continuous Wakefulness' *USAARL Technical Report No. 99–01*, US Army Aeromedical Research Laboratory, Fort Rucker, AL.

Caldwell, J.L., and Caldwell, J.A. (2013), 'Fatigue in Aviation,' In C. Kennedy and G. Kay, (Eds), *Aeromedical Psychology*, pp. 215–238. Ashgate Publishing Co., Farnham.

Caldwell, J.L., Caldwell, J.A., Colon, J., Ruyak, P.S., Ramspott, S., Sprenger, W.D., and Jones, R.W. (1998), 'Recovery of Sleep, Performance, and Mood Following 38 Hours of Sleep Deprivation using Naps as a Countermeasure,' *USAARL Report No. 98–37*, U.S. Army Aeromedical Research Laboratory, Fort Rucker, AL.

Caldwell J.L., Prazinko, B.F., Rowe, T., Norman, D., Hall, K.K., and Caldwell, J.A. (2003), 'Improving Daytime Sleep with Temazepam as a Countermeasure for Shift Lag,' *Aviation Space and Environmental Medicine*, Vol. 74(2), pp.153–63.

Carskadon, M.A., and Dement, W.C. (1987), 'Daytime Sleepiness: Quantification of a Behavioral State,' *Neuroscience and Biobehavioral Reviews*, Vol. 11(3), pp. 307–317.

Chellappa, S.L., Steiner, R., Blattner, P., Oelhafen, P., Götz,T., and Cajochen, C. (2011), 'Non-Visual Effects of Light on Melatonin, Alertness and Cognitive Performance: Can Blue-Enriched Light Keep Us Alert?' *PLoS ONE*, Vol. 6(1), e16429. Doi:10.1371/journal.pone.0016429.

Congressional Research Service (2002), 'Defense Budget for FY2003: Data Summary' In S. Daggett and A. Belasco (Eds), *CRS Report for Congress, RL31349*, Library of Congress, Washington, DC.

Cornum, R., Caldwell, J.A., and Cornum, C. (1997). 'Stimulant Use in Extended Flight Operations,' *Airpower Journal*, Spring, pp. 53–58.

Correll, J.T. (1998), 'Strung Out: We Have Too Few Forces and Too Little Money Chasing Too Many Open-Ended Deployments,' *Air Force Magazine Online*, Vol. 81(9), www.afa.org/magazine/editorial/09edit98.html (accessed 29 December 2015).

Dawson, D., and McCulloch. K. (2005), 'Managing Fatigue as an Integral Part of a Safety Management System,' *Proceedings of Fatigue Management in Transport Operations Conference*, Seattle, WA.

Dawson, D. and Reid, K. (1997), 'Fatigue, Alcohol and Performance Impairment' *Nature*, Vol. 388(6639), p. 235.

Department of the Air Force (1997), *Basic Air Force Doctrine Document 1*, US Air Force, Washington, DC.

Department of the Army (1991), 'Soldier Performance in Continuous Operations,' *FM 22–9*, Department of the Army, Washington DC.

Department of the Army (1996), 'Force of Decision: Capabilities for the 21st Century,' *White paper*, Department of the Army, Washington, DC.

Dinges, D.F. (1989), 'The Nature of Sleepiness: Causes, Contexts and Consequences,' In A. Stunkard and A. Baum (Eds) *Perspectives in Behavioral Medicine: Eating, Sleeping and Sex*, pp. 147–179, Hillsdale, NJ: Lawrence Erlbaum.

Dinges, D.F. (1990), 'The Nature of Subtle Fatigue Effects in Long-Haul Crews,' *Proceedings of the Flight Safety Foundation 43rd International Air Safety Seminar*, Flight Safety Foundation, Arlington, VA.

Dinges, D.F. (1995), 'An Overview of Sleepiness and Accidents,' *Journal of Sleep Research*, Vol. 4(Suppl. 2), pp. 4–14.

Dinges, D.F., Graeber, R.C., Rosekind, M.R., Samel, A. and Wemann, H.M. (1996), 'Principles and Guidelines for Duty and Rest Scheduling in Commercial Aviation,' *NASA Technical Memorandum No. 11040*, Ames Research Center, National Aeronautics and Space Administration, Moffett Field, CA.

Durmer, J.S., and Dinges, D.F. (2005). 'Neurocognitive Consequences of Sleep Deprivation, Sleep in Neurological Practice,' *Seminars in Neurology*, Vol. 25(1), pp. 117–129.

European Cockpit Association (2012), *Pilot Fatigue Barometer*, www.eurocockpit.be/sites/default/files/eca_barometer_on_pilot_fatigue_12_1107_f.pdf (accessed 12 December 2015).

Federal Aviation Administration (2011). *Fact Sheet – Pilot Fatigue Rule Comparison*, www.faa.gov/news/fact_sheets/news_story.cfm?newsId=13273 (accessed 29 December 2015).

Gander, P.H., Gregory, K.B., Graeber, R.C., Connell, L.J., Miller, D.L., and Rosekind, M.R. (1998a), 'Flight Crew Fatigue II: Short-Haul Fixed-Wing Air Transport Operations,' *Aviation, Space, and Environmental Medicine*, Vol. 69(9), Section II, pp. B8–B15.

Gander, P.H., Gregory, K.B., Miller, D.L., Graeber, R.C., Connell, L.J., and Rosekind, M.R. (1998b), 'Flight Crew Fatigue V: Long-Haul Air Transport Operations,' *Aviation, Space, and Environmental Medicine*, Vol. 69(9), Section II, pp. B37–B48.

Goode, J.H. (2003), 'Are Pilots at Risk of Accidents Due to Fatigue?' *Journal of Safety Research*, Vol. 34(3), pp. 309–313.

Grandner, M.A., Gallagher, R.A.L., and Gooneratne, N.S. (2013), 'The Use of Technology at Night: Impact on Sleep and Health,' *Journal of Clinical Sleep Medicine*, Vol. 9(2), pp. 1301–1302.

Hamelin P. (1987), 'Lorry Driver's Time Habits in Work and Their Involvement in Traffic Accidents,' *Ergonomics*, Vol. 30(9), pp. 1323–1333.

Harma, M. (1995), 'Sleepiness and Shiftwork: Individual Differences,' *Journal of Sleep Research*, Vol. 4(Supplement 2), pp. 57–61.

Harrington, J.M. (1994), 'Shift Work and Health – A Critical Review of the Literature on Working Hours,' *Annals of Academic Medicine, Singapore*, Vol. 23(5), pp. 699–705.

Hockey, R. (1983), *Stress and Fatigue in Human Performance*, John Wiley & Sons, New York.

Horne, J.A., and Östberg, O. (1976), 'A Self-Assessment Questionnaire to Determine Morningness-Eveningness in Human Circadian Rhythms,' *International Journal of Chronobiology*, Vol 4(2), pp. 97–110.

Hursh, S.R., Redmond, D.P., Johnson, M.L., Thorne, D.R., Belenky, G., Balkin, T.J., Storm, W.F., Miller, J.C., and Eddy, D.R. (2004), 'Fatigue Models for Applied Research in Warfighting,' *Aviation, Space and Environmental Medicine*, Vol. 75(Suppl), pp. A44–A53; discussion pp. A54–A60.

International Air Transport Association (IATA) (2014). Strong Demand for Air Travel Rises in 2014. file:///C:/Users/John/Documents/HP_Documents/IATA%20-%20Strong%20Demand%20for%20Air%20Travel%20Rises%20in%202014.html (accessed 29 December 2015).

International Civil Aviation Organization (2012), *Fatigue Risk Management Systems: Manual for Regulators, Document No. 9966*. Montréal, Quebec, Canada: ICAO, www.icao.int/safety/fatiguemanagement/frms%20tools/doc%209966%20-%20frms%20manual%20for%20regulators.pdf (accessed 29 December 2015).

Jackson, C.A. and Earl, L. (2006), 'Prevalence of Fatigue Among Commercial Pilots,' *Occupational Medicine*, Vol. 56(4), pp. 263–268,

Johns, M.W. (1991), 'A New Method for Measuring Daytime Sleepiness: The Epworth Scale,' *Sleep*, Vol. 14(6), pp. 540–545.

Kamimori, G.H., Johnson, D., Thorne, D., and Belenky, G. (2005), 'Multiple Caffeine Doses Maintain Vigilance During Early Morning Operations,' *Aviation, Space, and Environmental Medicine*, Vol. 76(11), pp. 1046–1050.

Keys, A., Brozek, J., Henschel, A., Mickelsen, O., and Taylor, H.L. (1950), *The Biology of Human Starvation*, 2 Vols., Minneapolis, MN: The University of Minnesota Press.

Kirsch, A.D. (1996), 'Report on the Statistical Methods Employed by the US FAA in Its Cost Benefit Analysis of the Proposed "Flight Crewmember Duty Period Limitations, Flight Time Limitations and Rest Requirements,"' *Comments of the Air Transport Association of America to FAA notice 95–18, FAA Docket No. 28081, Appendix D*, pp. 1–36.

Klein, D.E., Bruner, H., and Holtman, H. (1970), 'Circadian Rhythm of Pilot's Efficiency, and Effects of Multiple Time Zone Travel,' *Aerospace Medicine*, Vol. 41(2), pp. 125–132.

Kogi, K. (1985), 'Introduction to the Problems of Shiftwork,' In S. Folkard and T.H. Monk (Eds) *Hours of Work: Temporal Factors in Work Scheduling*, pp.165–184, John Wiley & Sons Ltd., Chichester.

Kolstad, J.L. (1989), 'National Transportation Safety Board Safety Recommendation,' *Evaluation of US Department of Transportation Efforts in the 1990s to Address Operator Fatigue, Appendix A, Report No. NTSB/SR-99/01*, pp 30–37, National Transportation Safety Board, Washington, DC.

Krause, K.S. (1999), 'Little Rock Aftermath,' *Trafficworld*, June, pp. 11–12.

Krueger, G.P. (1989), 'Sustaining Military Performance in Continuous Operations: Combatant Fatigue, Rest and Sleep Needs,' *Handbook of Military Psychology*, pp. 255–277, John Wiley and Sons, New York.

Landolt, H.P., Werth, E., Borbély, A.A., and Dijk, D.J. (1995), 'Caffeine Intake (200 mg) in the Morning Affects Human Sleep and EEG Power Spectra at Night,' *Brain Research*, Vol. 675(1–2), pp. 67–74.

Larter, D. (2014), 'America's military: The crushing deployment tempo'. *Military Times*, www.militarytimes.com/story/military/2014/12/14/americas-military-deployment-tempo-troops-families/20191377/ (accessed 6 May 2015).

Lauber, J.K., and Kayten, P.J. (1988), 'Sleepiness, Circadian Dysrhythmia, and Fatigue in Transportation System Accidents,' *Sleep*, Vol. 11(6), pp. 503–512.

Lerman, S.E., Eskin, E., Flower, D.J., George, E.C., Gerson, B, Hartenhaum, M., Hursh, S.R., and Moore-Ede, M. and ACOEM Presidential Task Force on Fatigue Risk Management (2012), 'Fatigue Risk Management in the Workplace,' *Journal of Occupational and Environmental Medicine*, Vol. 54(2), pp. 231–258.

Lewy, A.J., Bauer, V.K., Ahmed, S., Thomas, K.H., Cutler, N.L., Singer, C.M., Moffitt, M.T., and Sack, R.L. (1998), 'The Human Phase Response Curve (PRC) to Melatonin Is About 12 Hours Out of Phase with the PRC to Light,' *Chronobiology International*, Vol. 15(1), pp. 71–83.

Marcus, J.H., and Rosekind, M.R. (2015), 'Fatigue in Aviation: NTSB Findings and Safety Recommendations,' *Aerospace Medicine and Human Performance*, Vol. 86(3), p. 174.

Mitler, M.M., Carskadon, M.A., Czeisler, C.A., Dement, W.C., Dinges, D.F., and Graeber, R.C. (1988), 'Catastrophes, Sleep, and Public Policy: Consensus Report,' *Sleep*, Vol. 11(1), pp. 100–109.

Monk, T.H., and Folkard, S. (1985), 'Shiftwork and Performance,' *Hours of Work: Temporal Factors in Work Scheduling*, pp. 239–252, John Wiley & Sons, New York.

Moore-Ede, M. (1993), 'Aviation Safety and Pilot Error,' *Twenty-four Hour Society*, pp 81–95, Addison-Wesley Publishing Co., Reading, MA.

Moore-Ede M. (2009), 'Evolution of Fatigue Risk Management Systems: The "Tipping Point" of Employee Fatigue Mitigation,' CIRCADIAN White Papers, www.circadian.com/pages/157 white papers.cfm, accessed 15 April 2015.

Morisseau, D.S., and Persensky, J.J. (1994), 'A Human Factors Focus on Work Hours, Sleepiness and Accident Risk,' In T. Akerstedt and G. Kecklund (Eds) *Work Hours, Sleepiness and Accidents*, pp. 94–97, IPM and Karolinska Institute, Stockholm.

Musselman, B.T. (2008), *Human Factors. US Air Force Flying Safety Magazine*, January/February, Vol. 64(1–2), pp. 52–55.

Naitoh, P., Englund, C.E., and Ryman, D. (1982), 'Restorative Power of Naps in Designing Continuous Work Schedules,' *Journal of Human Ergology*, Vol. 11 (Suppl.), pp. 259–278.

National Sleep Foundation (2002), *Sleep in America Poll®: Adult Sleep Habits*, National Sleep Foundation, Washington, DC.

National Sleep Foundation (2008), *2008 Sleep in America® Poll: Sleep, Performance, and Workplace*, Washington, DC.

National Sleep Foundation (2012), *2012 Sleep in America® Poll: Planes, Trains, Automobiles and Sleep*, http://sleepfoundation.org/sleep-polls-data/sleep-in-america-poll/transportation-workers-and-sleep (accessed 29 December 2015).

National Sleep Foundation (2014), *2014 Sleep Health Index*, National Sleep Foundation, Arlington, VA.

National Sleep Foundation, 'The Sleep Environment,' www.sleepfoundation.org/article/how-sleep-works/the-sleep-environment (accessed 29 December 2015).

National Transportation Safety Board (1986), *China Airline Boeing 747-SP, N4522V, 300 Nautical miles northwest of San Francisco, California, February 19, 1985. NTSB/AAR-86–03*, National Transportation Safety Board, Washington, DC.

National Transportation Safety Board (1990), *Marine Accident Report-Grounding of the US Tankship Exxon Valdez on Bligh Reef, Prince William Sound, Near Valdez, Alaska, 24 Mar 1989. Report No. NTSB/Mar-90/04*, National Transportation Safety Board, Washington, DC.

National Transportation Safety Board (1994), *A Review of Flightcrew Involved, Major Accidents of US Air Carriers, 1978 through 1990. NTSB Safety Study No. SS-94–01*, National Transportation Safety Board, Washington, DC.

National Transportation Safety Board (2000), *Aircraft Accident Report: Controlled Flight into Terrain, Korean Air Flight 801, Boeing 747–300, HL7468, Nimitz Hill,*

Guam, August 6, 1997. Report No. NTSB/AAR-00/01, National Transportation Safety Board, Washington, DC.

National Transportation Safety Board (2001), Runway Overrun during Landing, American Airlines Flight 1420, McDonnell Douglas MD-82, N215AA, Little Rock, Arkansas, June 1, 1999. NTSB/AAR-01–02, National Transportation Safety Board. Washington, DC.

National Transportation Safety Board (2006), Collision with Trees and Crash Short of the Runway, Corporate Airlines Flight 5966 BAE Systems BAE-J3201, N875JX Kirksville, Missouri October 19, 2004. NTSB/AAR-06/01, National Transportation Safety Board, Washington, DC.

National Transportation Safety Board (2009), Loss of Control on Approach, Colgan Air, Inc., Operating as Continental Connection Flight 3407, Bombardier DHC-8–400, N200WQ, Clarence Center, New York, February 12, 2009. NTSB/AAR-10/01, National Transportation Safety Board, Washington, DC.

Office of Technology Assessment (1991), *Biological Rhythms: Implications for the Worker*, US Government Printing Office, Washington, DC.

Onen, S.H., Onen, F., Bailly, D., and Parquet, P. (1994), 'Prevention and Treatment of Sleep Disorders Through Regulation of Sleeping Habits,' *Presse Medicine*, Mar 12, Vol. 23(10), pp. 485–489.

Penetar, D., McCann, U., Thorne, D., Kamimori, G., Galinski, C., Sing, H., Thomas, M., and Belenky, G. (1993), 'Caffeine Reversal of Sleep Deprivation Effects on Alertness and Mood,' *Psychopharmacology*, Vol. 112(2–3), pp. 359–365.

Perry, I.C. (Ed.) (1974), 'Helicopter Aircrew Fatigue,' *AGARD Advisory Report No. 69*, Advisory Group for Aerospace Research and Development, Neuilly sur Seine, France.

Pigeau, R., Naitoh, P., Buguet, A., McCann, C., Baranski, J., Taylor, M., Thompson, M., and Mack, I. (1995), 'Modafinil, d-amphetamine, and Placebo during 64 Hours of Sustained Mental Work. Effects on Mood, Fatigue, Cognitive Performance and Body Temperature,' *Journal of Sleep Research*, Vol. 4(4), pp. 212–228.

Ramsey, C.S., and McGlohn, S.E. (1997), 'Zolpidem as a Fatigue Countermeasure,' *Aviation, Space, and Environmental Medicine*, Vol. 68(10), pp. 926–931.

Reyner, L.A., and Horne, J.A. (1997), 'Suppression of Sleepiness in Drivers: Combination of Caffeine with a Short Nap,' *Psychophysiology*, Vol. 34(6), pp. 721–725.

Reyner, L.A., and Horne, J.A. (1998), 'Evaluation of "In-Car" Countermeasures to Sleepiness: Cold Air and Radio,' *Sleep*, Vol. 4(1), pp. 46–50.

Ricci, J.A., Chee, E., Lorandeau, A.L., and Berger, J. (2007), 'Fatigue in the US Workforce: Prevalence and Implications for Lost Productive Work Time,' *Journal of Occupational and Environmental Medicine*, Vol. 49(1), pp. 1–10.

Roma, P.G., Hursh, S.R., Mead, A.M., and Nesthus, T.E. (2012), 'Flight Attendant Work/Rest Patterns, Alertness, and Performance Assessment: Field Validation of Biomathematical Fatigue Modeling,' *Report Number DOT/FAA/AM-12/12*. Federal Aviation Administration Office of Aerospace Medicine, Washington, DC.

Rosa, R.R., and Bonnet, M.H. (1993), 'Performance and Alertness on 8-Hour and 12-Hour Rotating Shifts at a Natural Gas Utility,' *Ergonomics*, Vol. 36(10), pp. 1177–1193.

Rosekind, M.R. (1994), 'Fatigue in Operational Settings: Examples from the Aviation Environment,' *Human Factors*, Vol. 36(2), pp. 327–338.

Rosekind, M.K., Co, E.L., Gregory, K.B., and Miller, D.L. (2000), 'Crew Factors in Flight Operations XIII: A Survey of Fatigue Factors in Corporate/Executive Aviation Operations.' *NASA/TM-2000–209610*, National Aeronautics and Space Administration, Ames Research Center, Moffett Field, CA.

Rosekind, M.R., Graeber, R.C., Dinges, D.F., Connell, L.J., Rountree, M.S., Spinweber, C.L., and Gillen, K.A. (1994), 'Crew Factors in Flight Operations IX: Effects of Planned Cockpit Rest on Crew Performance and Alertness in Long-Haul Operations,' *NASA Technical Memorandum no. 108839*, National Aeronautics and Space Administration, Ames Research Center, Moffett Field, CA.

Rupp, T.L., Wesensten, N.J., Bliese, P.D., and Balkin, T.J. (2009), 'Banking Sleep: Realization of Benefits During Subsequent Sleep Restriction and Recovery,' *Sleep*, Vol. 32(3), pp. 311–321.

Samel, A., Wegmann, H-M. and Vejvoda, M. (1997), 'Aircrew Fatigue in Long-Haul Operations,' *Accident Analysis and Prevention*, Vol. 29(4), pp. 439–452.

Samel, A., Wegmann, H-M., Vejvoda, M., Drescher, J., Gundel, A., Manzey, D., and Wenzel, J. (1997), 'Two-Crew Operations: Stress and Fatigue during Long-Haul Night Flights,' *Aviation, Space, and Environmental Medicine*, Vol. 68(8), pp. 679–687.

Scott, A.J. (1990), 'Shiftwork,' *Occupation Medicine: State of the Art Reviews*, Vol. 5(2), Hanley & Belfus, Inc., Philadelphia.

Shapiro, C.M., and Heslegrave, R.J. (1996), *Making the Shift Work*, Joli Joco Publications, Toronto.

Stevens, R.G., Hansen, J., Costa, G., Haus, E., Kauppinen, T., Aronson, K.J., Castaño-Vinyals, G., Davis, S., Frings-Drewsen, M.H.W., Fritschi, L., Kogevinas, M., Kogi, K., Lie, J-A., Lowden, A., Peplonska, B., Pesch, B., Pkkala,E., Schernhammer, E., Travis, R.C., Vermeulen, R., Zheng, T., Cogliano, V., and Straif, K. (2011), 'Considerations of Circadian Impact for Defining "Shift Work" in Cancer Studies: IARC Working Group Report,' *Occupation and Environmental Medicine*, Vol. 68(2), pp. 154–162.

Talillard, J., Capelli, A., Saagaspe, P., Anund, A., Akerstedt, T., and Philip, P. (2012), 'In-Car Nocturnal Blue Light Exposure Improves Motorway Driving: A Randomized Controlled Trial,' *PLOS ONE*, Vol. 7(10), e46750. Doi:10.1371/journal.pone.0046750.

Tilghman, A. (2014), America's military: Readiness on a shoestring. *Military Times*, www.militarytimes.com/story/military/2014/12/13/americas-military-readiness-on-a-shoestring/20186051/ (accessed 13 December 2015).

Van Dongen, H.P.A. (2004). 'Comparison of Mathematical Model Predictions to Experimental Data of Fatigue and Performance,' *Aviation, Space, and Environmental Medicine*, Vol. 73(3, Suppl.), pp. A15–A36.

Van Dongen, H.P.A., Maislin, G., Mullington, J.M., and Dinges, D.F. (2003), 'The Cumulative Cost of Additional Wakefulness: Dose-Response Effects on Neurobehavioral Functions and Sleep Physiology From Chronic Sleep Restriction and Total Sleep Deprivation,' *Sleep*, Vol. 26(2), pp. 117–126.

Vandewalle, G., Maquet, P., and Dijk, D.-J. (2009), 'Light as a Modulator of Cognitive Brain Function,' *Trends in Cognitive Science*, Vol. 13(10), pp. 429–438.

Viola, A.U., James, L.M., Schlangen, L.J.M., and Dijk, D.-J. (2008), 'Blue-enriched White Light in the Workplace Improves Self-reported Alertness, Performance and Sleep Quality,' *Scandinavian Journal of Work and Environmental Health*, Vol. 34(4), pp. 297–306.

Watson, N.R., Badr, M.S., Belenky, G., Bliwise, D.L., Buxton, O.M., Buysse, D., Dinges, D.F., Gangwisch, J., Grandner, M.A., Kushida, C., Malhotra, R.K., Martin, J.L., Patel, S.R., Quan, S.F., and Tasali, E. (2015), 'Recommended Amount of Sleep for a Healthy Adult: A Joint Consensus Statement of the American Academy of Sleep Medicine and Sleep Research Society,' *Sleep*, Vol. 38(6), pp. 843–844.

Webb, W.B. (1982), *Biological Rhythms, Sleep, and Performance*, John Wiley & Sons, New York.

Wesensten, N.J., Killgore, W.D.S., and Balkin, T.J. (2005), 'Performance and Alertness Effects of Caffeine, Dextroamphetamine, and Modafinil During Sleep Deprivation,' *Journal of Sleep Research*. Vol. 14(3), pp. 255–266.

Wright, N., and McGown, A. (2001), 'Vigilance on the Civil Flight Deck: Incidence of Sleepiness and Sleep during Long-Haul Flights and Associated Changes in Physiological Parameters,' *Ergonomics*, Vol. 44(1), pp. 82–106.

Index

(Figures indexed with **bold** page numbers and Tables indexed with *italic* page numbers)